Series/Number 07-099

UNIVARIATE TESTS
FOR TIME SERIES MODELS

JEFF B. CROMWELL
West Virginia University

WALTER C. LABYS
West Virginia University

MICHEL TERRAZA
University of Montpellier I, France.

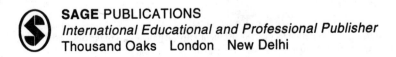
SAGE PUBLICATIONS
International Educational and Professional Publisher
Thousand Oaks   London   New Delhi

For information address:

SAGE Publications, Inc.
2455 Teller Road
Thousand Oaks, California 91320
E-mail: order@sagepub.com

SAGE Publications Ltd.
6 Bonhill Street
London EC2A 4PU
United Kingdom

SAGE Publications India Pvt. Ltd.
M-32 Market
Greater Kailash I
New Delhi 110 048 India

Printed in the United States of America

Library of Congress Catalog Card No. 89-043409

Cromwell, Jeff B.
    Univariate tests for time series models / Jeff B. Cromwell, Walter
C. Labys, Michel Terraza.
        p.   cm.—(Quantitative applications in the social sciences; 99)
    Includes bibliographical references.
    ISBN 0-8039-4991-X (pb)
    1. Social sciences—Statistical methods.   2. Time-series analysis.
I. Labys, Walter C., 1937-   . II. Terraza, Michel. III. Title.
IV. Series: Sage university papers series. Quantitative
applications in the social sciences; no. 99.
HA30.3.C76   1994
300′.1′51955—dc20   93-36737

96 97 98 99 00 01 10 9 8 7 6 5 4 3 2

Sage Production Editor: Astrid Virding

When citing a university paper, please use the proper form. Remember to cite the current Sage University Paper series title and include the paper number. One of the following formats can be adapted (depending on the style manual used):

(1) CROMWELL, JEFF B., LABYS, WALTER C., & TERRAZA MICHEL (1994) Univariate Tests for Time Series Models. Sage University Paper series on Quantitative Applications in the Social Sciences, 07-099. Thousand Oaks, CA: Sage.

OR

(2) Cromwell, J. B., Labys, W. C., & Terraza, M. (1994). *Univariate tests for time series models* (Sage University Paper series on Quantitative Applications in the Social Sciences, series no. 07-099). Thousand Oaks, CA: Sage.

CONTENTS

SERIES EDITOR'S INTRODUCTION

Time series data are different, in that observations on the unit of analysis have a temporal order. When a time series variable Y_t is plotted over, say, 100 months, it may show off an intriguing pattern. We could try to account for that pattern by introducing outside explanatory variables, or by examining the history of the variable itself. The latter approach leads to time series modeling, aimed at uncovering an internal process responsible for the variable's behavior. The first, critical step in time series modeling is identification. Fortunately, there are many identification tests now available, and they are ably explicated in the monograph at hand.

Professors Cromwell, Labys, and Terraza neatly map out an identification strategy, clearly describing the necessary tests at each stage. Here is a sketch, with some illustrative tests. Initial tests (e.g., the Dickey-Fuller) are made for stationarity. (The authors point out that stationarity must be established, before modeling can go forward.) Then, normality tests (e.g., the Jarque-Bera) are performed, as are tests for independence (e.g., the widely used Box-Pierce). Most time series do not show independence, so the testing would go on to sort for linear versus nonlinear dependence (e.g., Keenan). If the dependence is linear, as commonly assumed, then the question is whether the process is autoregressive (AR), moving average (MA), or some combination, (e.g., ARMA). (If it is nonlinear, then another specification, e.g., autoregressive conditional heteroscedastic model—ARCH—might be preferred.) After this specification, it is necessary to test for lag order. For example, a Likelihood Ratio test may indicate the model is AR(1), rather then AR(2). Finally, the results of this identification procedure are validated (or not) through the examination of residuals.

Of course, at each stage in the identification process, more than one test is possible. For instance, with regard to tests for independence, the authors cover the Ljung-Box and Box-Pierce; Turning Point; Runs; Rank Version of von Neumann; and the Brock, Dechert, and Scheinkman. Further, they evaluate the relative merits of the tests, exhibiting sensitivity to the "judgment calls" that are sometimes necessary. As a case in point, they observe that, unless the goal is to estimate "pure" AR and MA models, it "may be a good idea to use all the measures mentioned" before deciding on model order.

The authors are careful to provide empirical examples. Indeed, throughout they use the same data—a time series on macro-political partisanship in the United States—to illustrate the tests. This provides continuity and, as well, nicely shows how test decisions can influence results. To help the reader actually carry out the tests, available software is reviewed, in a systematic comparison of three programs—MicroTSP, RATS, SHAZAM. Without doubt, this exhaustive handbook of univariate tests for time series models will stand unrivaled for some time.

—*Michael S. Lewis-Beck*
Series Editor

UNIVARIATE TESTS
FOR TIME SERIES MODELS

JEFF B. CROMWELL
West Virginia University

WALTER C. LABYS
West Virginia University

MICHEL TERRAZA
University of Montpellier I, France

1. INTRODUCTION

Time series modeling continues to be a major interest of social scientists. These models have the advantage that behavioral patterns potentially can be explained or predicted simply by studying the past history of a variable reflecting those patterns. More complex models can be constructed by examining and by employing the interrelationships between several such variables. The most important aspect of constructing these models is learning about the intrinsic time patterns of a variable or its underlying generating process.

Only then can a time series model be selected that will replicate this process. Unfortunately the task of deciphering the generating process is a difficult one. In fact Granger and Newbold (1986) among others assert that the most difficult stage of model building is the first or identification stage. Figure 1.1 shows the additional stages, which include specification, estimation, and diagnostic checking. Historically,

AUTHORS' NOTE: We would like to thank the following individuals for their comments on previous portions of this manuscript: Jim Granato, Montague Lord, Veronique Murcia, David Sorenson, Jacqueline Khorassini, Douglas Mitchell, Chen Lin, Fujimoto Hiroaki, and Yongjie Hu. Our appreciation is also due to graduate students in the Econometrics Seminar at West Virginia University and students in the Time Series Seminar at the Inter-University Consortium for Political and Social Science Research at the University of Michigan. We would especially like to thank Michael Lewis-Beck, Clive Granger, Henry Heitowit, and Virgil Norton for their advice and encouragement.

1

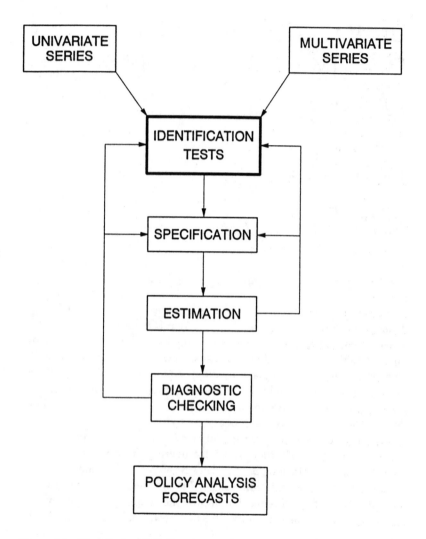

Figure 1.1. Time Series Modeling Approach

only a few time series tests have been available to enable researchers to choose from a wide class of models a single one that might adequately describe a given time series pattern. Today, however, a much greater number of identification tests are available (e.g., see Labys, Murcia, &

Terraza, 1991). Although the existence of these tests enables us to be better modelers, a larger task meets us in attempting to decide which of several test methods, in fact, should be employed.

The purpose of this monograph is to provide practical guidelines for social scientists concerning the nature of these increasingly complex tests and their application in identifying univariate time series models. Methods for identifying multivariate models will appear in a second monograph, *Multivariate Tests for Time Series Models*. Our goal has been to present the most popular tests employed in the time series literature together with related testing software, so as to make further applications helpful and meaningful.

Definition

The definition of a time series begins with the notion of a stochastic or random process, which is defined as an ordered collection of random variables indexed by time $x(t_1)$, $x(t_2)$, . . . , $x(t_n)$. Time series data that can be employed by social scientists and thus described by such variables tend to be discrete, that is, daily, monthly, yearly, and so forth. Consider, for example, the sample time series or realization of a stochastic process shown in Figure 1.2. This figure represents a plot of quarterly macropartisanship data on the party identification of voters beginning in January 1953 and extending to December 1988. These data have been estimated by Stimson (1991), who appropriately combined several major survey series. The interesting properties discovered for this series by MacKuen, Erikson, and Stimson (1992) makes them appropriate for the test examples presented throughout this monograph.

Figure 1.2 provides a typical visual representation of a time series with the Y axis defined as the random variable, $D(t)$, describing party identification, and the X axis as the time index, which is measured by integer values sampled at quarterly intervals of time. In order to make the connection between the sample $D(t)$ and the set of all samples $\{D(t)\}$, consider the above time series. The quarterly measure is either the percentage of voters who are Democrat on the last month of the quarter, first month, or one of the other two months, or it could possibly be an average of all the months over that measured quarter. If the average is used, then we expect it to be a good estimate for the quarter. Thus the measured variable for the first quarter, denoted by $D(1)$, could represent the mean value of all the set of all possible values for the quarter, $\{D(1)\}$.

4

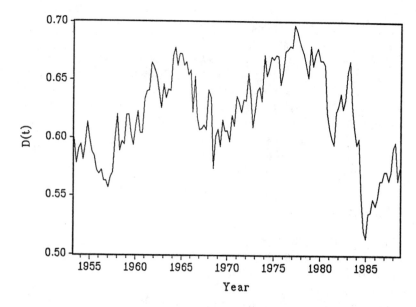

Figure 1.2. Macropartisanship [Percentage of voters who are democrats, $D(t)$]

Independence

An important characteristic of Figure 1.2 is the existence of some type of regularity. The most fundamental characteristic that differentiates time series data from other types of data is that, in general, the values of a time series at different time instants or points, $x(t_1)$ and $x(t_2)$, will be related. For example, it appears that party identification has grown from 1952 until 1978, after which a decline begins. At the same time one can detect a distinct short-term cycle imposed on these trends.

If you extend this idea to more than 2 points in time, the random variables, $x(t_i)$ and $x(t_j)$, for T observations with $i, j = 1, \ldots, T$ can be either statistically dependent or independent. *Statistical independence* requires that the joint probability distribution of a time series, a collection of T random variables, can be written as a product of marginal distributions for each of the T random variables, for example, $F[x(1), x(2), \ldots, x(T)] = f[x(1)]f[x(2)] \ldots f[x(T)]$. In practice, it is quite often the case that independence is examined through the moments of the joint distribution. For example in the case where $T = 2$, if the expected value of two random variables is the product of the expected value of each random variable, $E[x(1)x(2)] = E[x(1)]E[x(2)]$, then the two variables are *partially* independent and also *uncorrelated* in the case where $x(t)$ has a finite variance. Otherwise, if equality does not hold, then dependence results. This, of course, can be extended to the case of T random variables or any general time series. Due to its critical importance, one must be cautious about invoking the independence assumption in time series analysis.

It is further important to make the distinction between linear dependence and nonlinear dependence. Linear or serial dependence results in the case that the random variables $x(t)$ and $x(s)$ are statistically dependent for at least some pairs of values of t, s for $t \neq s$. The case of serial dependence in the time series literature is known as *correlation*. Given that some dependence exists in the data, it is the task of the time series analyst to examine the intrinsic nature of the type and form of dependency between values at different time points, and to try to construct statistical models that can reproduce this dependency. Once the type and form of dependency are determined, time series modeling then involves the formulation of mathematical models that can emulate the underlying generating process inherent in a time series. Once this process is completed, then one can use the model to analyze particular policies or to forecast.

Decomposition

As a way of classifying dependence, a time series can be decomposed into three components: trend, cyclical, and irregular. The first two components are typically referred to as the *deterministic* components. Because the normal procedure is to remove or to filter out these components, it is important to understand them conceptually. The trend component appears in the form of a change in the mean value and/or variance of the series. For example, a plot of a time series could be linear. That is, the mean of the series could be increasing or decreasing, such as in a linear trend. However, all trends do not have to be linear; they can also be linear or nonlinear in the form of higher polynomials or exponential.

The cyclical component can be associated with formal cyclical models such as sinusoids and their more complex harmonic patterns, that is, combinations of sine and cosine functions. The cyclical component can also be ascribed to seasonal variation in the time series, that is, variations in quarterly or monthly retail sales could be explained by the seasonal effect. The irregular component is what is left after the trend and cyclical including seasonal components have been removed (e.g., see Terraza, 1980). Our principal interest here is with the irregular component and with the selection of probabilistic models that can describe the time series dependency in the irregular component.

Structure

Given the existence of some type of dependency in the irregular component, analyzing the structure of a time series requires the use of moments as defined in (1.1). The first moment, $\mu(t)$, is a measure of location or mean; the second central moment, $\sigma^2(t)$, is a measure of dispersion or variance. A measure of the strength of the relationship between $x(r)$ and $x(s)$ in the form of the covariance, $\tau_t(r, s)$, also is needed.

(i) $\mu(t) = E[x(t)]$

(ii) $\sigma^2(t) = \text{var}[x(t)] = E\{[x(t) - \mu]^2\}$

(iii) $\tau_t(r, s) = \text{cov}[x(r), x(s)] = E\{[x(r) - \mu_r][x(s) - \mu_s]\}$ (1.1)

where $t, r, s = 1, \ldots, T$. Since in this case, the covariance is restricted to just one variable $x(t)$, it is called the *autocovariance* and measures the dependence between two values of the time series observed at different points in time. Note that the covariance measure depends only on first and second moments of $x(t)$. One deficiency of the autocovariance measure is that the value of the expression in (iii) depends on the value of $x(r)$ and $x(s)$. A standardization can be performed to achieve a measure of *autocorrelation* by dividing the covariance in (iii) by the product of the standard deviation of $x(r)$ and $x(s)$

$$\rho_t(r, s) = \tau_t(r, s)/[\tau_t(r, r)\tau_t(s, s)]^{\frac{1}{2}} \qquad (1.2)$$

where $|\rho_t(r, s)| < 1$.

Stationarity

A most important property of a time series is that it be *strictly stationary*. This requires that the sample statistics defined in (1.1) and the resulting standardization in (1.2) are all invariant to any time shifts. For a strictly stationary process $x(t)$, the autocovariance can be expressed as

$$\tau(k) = E\{[x(t) - \mu][x(t + k) - \mu]\} \tag{1.3}$$

This is known as the autocovariance coefficient at lag k. Notice that because the mean is assumed to be constant, the difference between $x(t)$ and $x(t + k)$ will depend only on the distance between t and $t + k$. This expression also requires performing the standardization

$$\rho(k) = \tau(k)/\tau(0) = \frac{E\{[x(t) - \mu][x(t + k) - \mu]\}}{E\{[x(t) - \mu]^2\}} \tag{1.4}$$

where $\tau(0)$ represents the variance of the process and $\rho(k)$ is commonly referred to as the *autocorrelation function (ACF) at lag* k.

Although the property of stationarity need not be present in order to define the autocorrelation function, the interpretation of that function is improved in the presence of strict stationarity. Throughout this study, we will say weakly stationary or stationary in the wide sense when the mean and variance are constant but not higher orders (i.e., skewness, kurtosis, etc.), and strictly stationary when the process is both stationary of order two and normally distributed. For correct interpretation of the above measures of (1.1) and (1.2), one only needs the assumption of a weakly stationary time series.

It is also important to understand some of the properties of the autocorrelation function for stationary processes. For example, when $x(t)$ is a stationary and mean-zero process, then

$$\text{(i)} \quad \rho(k) = \frac{E[x(t)x(t + k)]}{E[x^2(t)]} \tag{1.5}$$

$$\text{(ii)} \quad \rho(0) = \frac{E[x(t)x(t)]}{E[x^2(t)]} = 1$$

$$\text{(iii)} \quad |\rho(k)| < 1 \text{ implies } |E[x(t)x(t + k)]| < E[x^2(t)]$$

Condition (iii) is important to remember, because it implies that the absolute value of the correlation at lag k is bounded by the variance of the process $x(t)$. These properties will be used in certain test developments of subsequent chapters.

Time Series Tests and Model Building

The purpose of this monograph is to explain how a battery of statistical tests can be applied to identify which kinds of univariate time series models can be constructed and applied. The model building procedure adopted here is a sequential one based on the above identification tests, as shown in Figure 1.3. Although this procedure appears to be fairly mechanical, a good deal of judgment is required in reaching a final model specification. The first of these sequential steps requires testing for stationarity in the variable or process under scrutiny. The second step involves testing for normality to provide evidence of the type of dependence of a stationary process. The third step requires determining whether the same series is, in fact, statistically independent. If the null hypothesis of independence is rejected, then the alternative hypothesis implies the existence of linear or nonlinear dependence. The test of this hypothesis occurs at the fourth step. If the data are linearly dependent, then linear time series models must be specified. Alternatively, if the data are nonlinearly dependent, then nonlinear time series models must be employed.

The nonlinear model specification process begins with the fifth step, in which either additive or multiplicative nonlinear models are constructed. For example, in the case where nonlinearity exists in the mean of the process, then a bilinear, threshold, or an exponential model can be used. Alternatively, if the nonlinearity enters through the variance of the process, then an ARCH or GARCH model can be used. It should be clear that the model specification process prescribed here is primarily statistical in nature. Model specification also depends on how particular economic, political, or other theories might be embodied in a particular model and appropriately tested. Such a model selection process has recently been advanced by Granger, King, and White (1992).

Once a particular linear or nonlinear model is selected, the next step is to decide on the number of lags to be included in the model based on various lag order determination tests. Once the order is determined and the assumptions concerning the residuals are verified, then the modeling

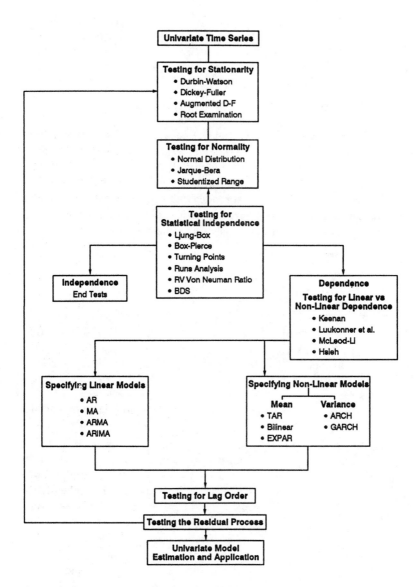

Figure 1.3. Univariate Test Procedure for Model Identification

process is complete. Verification of these assumptions can be achieved through validation tests that are based on the diagnostic checking of model residuals or other measures of fit.

2. TESTING FOR STATIONARITY

Stationarity is the first fundamental statistical property tested for in time series analysis, because most statistical models require that the underlying generating processes be stationary. If a process is not stationary (nonstationary), then hopefully a transformation of the data can be performed in such a way as to render that process weakly stationary. Because most real world data are nonstationary, performing transformations and testing for stationarity should coincide.

Transformations

As a first step, we should have some idea of the kinds of data transformation that can change a nonstationary time series into a stationary one. The most common transformation is that of differencing, that is, subtracting a past value of a variable from its current value. For example, one can remove the effect of a linear trend in a time series by performing a first difference. In this case, using the transformed variable $\Delta x(t) = x(t) - x(t - 1)$ is sufficient to remove a linear trend. One useful way of characterizing this difference operation is through the use of the lag operator, L. For example, the simplest operation is $Lx(t) = x(t - 1)$. Higher powers of L result in longer lags, $L^2 x(t) = x(t - 2)$, and so on. Therefore, it is easily seen that applying operator L to $x(t)$ in the following format results in $[1 - L]x(t) = x(t) - x(t - 1)$. Higher order difference transformations can be accomplished by taking higher differences, d, of the term in brackets, $[1 - L]^d$. For example, to remove a quadratic trend, execute the transform $[1 - L]^2 x(t) = x(t) - 2x(t - 1) + x(t - 2)$. These difference transformations act like a *filter* to remove linear and polynomial trend components. The use of differencing to render a time series stationary was popularized by Box and Jenkins (1976) in the context of building linear integrated ARMA models, more popularly known as ARIMA.

As stated previously, second order or weak stationarity requires that a process must have a constant variance along with a constant mean. If nonstationarity enters through the variance of a process, then heteroscedasticity (i.e., a nonconstant variance) becomes a problem. One way to counteract this effect is by performing a power transformation on the sample time series $x(t)$. Power transformations enable one to render a time varying variance constant and/or render a nonnormal process normal. Probably the most commonly used group of power transformations is the Box-Cox family of power transformations that are defined by

$$\varphi_1[x(t)] = [x^{(\lambda)}(t) - 1]/\lambda$$

For example, if the variance of this series, $x(t)$, is dependent on time and the conditional variance is log-normal, then it can be shown that a logarithmic transformation such as $\lambda = 0$ will render the variance constant, $[x^{(0)}(t) - 1]/\lambda = \log x(t)$. A log transformation is also appropriate when the standard deviation is believed to be directly proportional to the mean.

In other cases, other types of power transformations can be performed to accomplish a similar result. For example, if a variable, $x(t)$, is bounded in the unit interval, [0,1], Wallis (1987) suggests using the logistic transformation

$$\varphi_2[x(t)] = \ln\{x(t)/[1 - x(t)]\}$$

When the mean and variance both trend together, Granger and Hughes (1971) have suggested the following transformation

$$\varphi_3[x(t)] = x(t)/(2m + 1)^{-1}\sum_j x(t - j)$$

where $j = -m$ to m.

Of course, both the log transformation and the differencing transformation can be performed together. For example, if an initial log transformation is followed by first differencing, then the time series is changed from levels (raw data) to growth rates. Figure 2.1 illustrates this concept by showing the variable $D(t)$ featured in Figure 1.2 transformed into growth rates, $[1 - L]\log[D(t)]$. As stated above, it has become quite commonplace to use first differences of logs before starting to analyze a series. However, as will be pointed out in subsequent sections, this a priori transforming may not always be desirable.

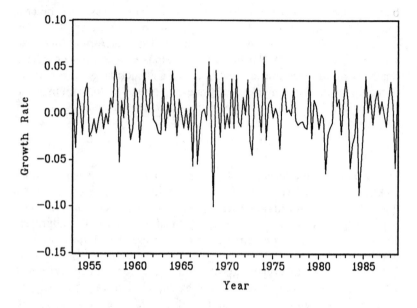

Figure 2.1. Growth Rates of Macropartisanship [Percentage changes in percentage of voters who are democrats, $D(t)$]

Random Walk

Suppose that a difference operator $\Delta x(t) = x(t) - \alpha x(t-1)$ is employed, where $\alpha = 1$. If we add an independent and stationary process, $e(t)$, to this operator and set this expression to zero, then we obtain $x(t) = x(t-1) + e(t)$ or $\Delta x(t) = e(t)$. As is discussed in Chapter 6, the specification of this *model* requires that $|\alpha| < 1$ in order for the process $x(t)$ to be stationary; however, if $\alpha = 1$, then the process is nonstationary and is said to possess a *unit root*, that is, it is integrated of order one, I(1), where $\Delta x(t) = x(t) - x(t-1)$. Higher orders of integration d, or more than one unit root involve the previously mentioned operator $[1 - L]^d$. It can be seen for the specification of $x(t) - x(t-1) = e(t)$ or $x(t) = x(t-1) + e(t)$ that $x(t)$ has one unit root and is called a *random walk* process. Time series of this type can be made stationary by performing a first order difference transformation. This implies that we would want to difference a time series only in the presence of one or multiple unit roots. A unit root implies that the trend component of the data is stochastic. It is quite possible for the trend component to be nonstochastic; this implies what is called *trend stationary* as opposed to difference stationary with the unit root concept.

This distinction is important because, if a time series is differenced even though it does not possess a unit root, a unit root is introduced in the residual process (i.e., moving average) of the time series. This outcome is referred to as the problem of overdifferencing. For example, suppose that $x(t) = e(t) - \beta_1 e(t - 1)$, which is commonly referred to as an MA(1) process. If we difference $x(t)$ we have

$$x(t) - x(t - 1) = e(t) - \beta_1 e(t - 1) - [e(t - 1) - \beta_1 e(t - 2)]$$
$$= e(t) - (1 + \beta_1)e(t - 1) + \beta_1 e(t - 2)$$

Although the process is still stationary, it is no longer invertible because the parameters on the residuals now sum to one. Thus, a unit root is introduced in the moving-average component of the process. It can also be shown that the variance of the overdifferenced process will be larger than that of the original process. Anderson (1971) pointed out that one can use the following rule of thumb for determining the degree of differencing. Because the sample variance will decrease until a stationary process is found and will increase in the case of overdifferencing, one need only to keep differencing until the sample variance increases. However, as Mills (1990) points out, this is not always the case. So other methods must be employed.

If the process is trend stationary, then the trend component can be removed from the time series by including a time index into a regression equation. Although testing for trend stationary as opposed to difference stationary is quite difficult, Cochrane (1988) provides a measure to serve as a guide for making the distinction.

Because random walk processes are also nonstationary, it seems that testing the value of α in the equation, $x(t) - \alpha x(t - 1) = e(t)$ might enable us to determine if the time series, $x(t)$, behaves like a random walk.

Dickey-Fuller (DF) Test

The Dickey-Fuller test examines the condition where the process has a unit root and where differencing helps to remove this root. This test requires performing the regression

$$\Delta x(t) = \alpha_1 x(t - 1) + e(t) \qquad (2.1)$$

where $e(t)$ is an independent and stationary process. The test involves examining the t statistic for the parameter α_1, under the null hypothesis that $\alpha_1 = 0$. The null hypothesis is that $x(t)$ is nonstationary; this implies that $x(t)$ is a random walk, that is, it possesses a unit root. The alternative hypothesis reflects the fact that $\alpha_1 \neq 0$. The null hypothesis is rejected if the t statistic is larger than the critical value, τ_1, obtained from Appendix Table A.1 or from MacKinnon (1990). Note that the critical values are not the ones used in typical applications of the t statistic, $t(\alpha_1)$, in linear regression models. Rather they consist of values estimated in Monte Carlo simulations of the test statistic performed by Dickey and Fuller (1979) and MacKinnon (1990).

It is possible that the time series could also behave as a random walk with a drift. This means that the value of $x(t)$ may not center to zero [i.e., $x(t)$ has a non-zero mean] and thus a constant should be added to the DF equation (2.1). In this case, the critical value of the statistic becomes τ_2 and the null hypothesis is rejected when $t(\alpha_1) > \tau_2$. A linear trend variable could also be added along with the constant to this equation, which results in the null hypothesis reflecting stationary deviations from a trend. This would be a combination of both trend and difference stationary components. Rejection of the null hypothesis in this situation occurs when $t(\alpha_1) > \tau_3$. The critical values for these modifications are also repeated in Appendix Table A.1. MacKinnon (1990) also provides critical values for these three options of neither constant nor trend included, only constant included, and both constant and trend included. These critical values are based on more replications than those used to generate Appendix Table A.1, which suggests that the former may be more accurate. One can see that the power of these tests will depend on the assumption that $e(t)$ is an independent, stationary process. Selection of τ_1, τ_2, or τ_3 depends on the choice of which regression equation provides stronger evidence that $e(t)$ in any of the equations is an independent and stationary process.

Test Procedure

Step 1. The Dickey-Fuller test requires that several test regressions are specified and estimated to test the null hypothesis that $x(t)$ is a random walk in one of three ways, depending on the choice of an intercept and/or a trend term.

$$(i) \ \Delta x(t) = \alpha_1 x(t-1) + e(t)$$

(ii) $\Delta x(t) = \alpha_0 + \alpha_1 x(t-1) + e(t)$

(iii) $\Delta x(t) = \alpha_0 + \alpha_1 x(t-1) + \beta t + e(t)$

The null hypothesis for each is

(i) H_0: $x(t)$ is a random walk, $\alpha_1 = 0$
(ii) H_0: $x(t)$ is a random walk plus drift, $\alpha_1 = 0$
(iii) H_0: $x(t)$ is a random walk plus drift around a stochastic trend, $\alpha_1 = 0$

The associated alternative hypothesis for each null is two sided, that is H_A: $\alpha_1 \neq 0$.

Step 2. For each case i, calculate the test statistic

$$t_i(\alpha_1) = [(\alpha_1 - 0)/\text{SE}(\alpha_1)]$$

where $\text{SE}(\alpha_1)$ refers to the standard error of α_1 .

Step 3. Choose the significance level, α, and critical values, τ_i, for each i from Appendix A.1 and/or MacKinnon (1990).

Step 4. Reject the null hypothesis if

$$|t_i(\alpha_1)| > |\tau_i|$$

Example

Step 1. The DF test regressions are estimated for the party identification variable of Figure 1.2 over the sample period 1953.2-1988.4.

(i) $\Delta D(t) = -0.0006487 D(t-1) + e(t)$

(ii) $\Delta D(t) = 0.05431 - 0.087734 D(t-1) + e(t)$

(iii) $\Delta D(t) = 0.05553 - 0.08747 D(t-1) - 0.000018t + e(t)$

Step 2. The computed $t_i(\alpha_1)$ statistics for α_1 are

(i) $t_1(\alpha_1) = (-0.000649 - 0)/0.0023361 = -0.2777$

(ii) $t_2(\alpha_1) = (-0.087734 - 0)/0.0353481 = -2.4820$

(iii) $t_3(\alpha_1) = (-0.087475 - 0)/0.0354415 = -2.4680$

Step 3. The corresponding critical values for a significance level of $\alpha = 0.05$ based on MacKinnon (1990) are

(i) $\tau_1 = -1.9421$

(ii) $\tau_2 = -2.8818$

(iii) $\tau_3 = -3.4419$

Step 4. Because the regression values of t are less than the critical value, τ_i, the null hypothesis is accepted that $x(t)$ is nonstationary or, in this case, possesses a unit root and is a random walk for all three null hypotheses.

Augmented Dickey-Fuller (ADF) Test

It is quite probable that the assumption of $e(t)$ being independent will not be realized (i.e., because of the existence of serial correlation) under the cases presented under the Dickey-Fuller test. Because of this probability, higher order lags may be necessary to remove the serial correlation. The DF test regression (2.1) can be augmented accordingly as follows

$$\Delta x(t) = \alpha_1 x(t-1) + \sum_j \beta_j \Delta x(t-j) + e(t) \quad \text{for } j = 1, 2, \ldots, p \quad (2.2)$$

where $e(t)$ is an independent, stationary process, and p is the number of lags chosen for the dependent variable, $\Delta x(t)$. Once again the null hypothesis is that $\alpha_1 = 0$, that is, $x(t)$ has a unit root. The specification of the lag structure of (2.2) enables the ADF test to account for a more dynamic specification of the regression than the DF test. However, the number of lags, p, must be selected empirically. The reader is referred to Chapter 8 for a discussion on empirical lag determination. Once the order p is selected, then the t statistic of α_1 and the critical values can be examined. Critical values based on the work of Dickey and Fuller have been estimated for any case. The MacKinnon (1990) critical values discussed above (which are employed in the software package MicroTSP 7.0) are valid for any value of p; they also vary with the three test options as well as the number of observations. We particularly recommend using the MacKinnon approach.

The procedure for rejecting the null is exactly the same as in the case of the Dickey-Fuller test. We advise using MicroTSP 7.0, which also allows the same three options of including a constant and/or trend term in the specification of (2.2).

Test Procedure

Step 1. The test regressions of the DF test are now performed with the inclusion of a lag representation. Lag order, p, is selected based on the tests in Chapter 8. For cases (i)-(iii) the summation is from $j = 1$ to p.

$$\text{(i) } \Delta x(t) = \alpha_1 x(t-1) + \sum_j \beta_j \Delta x(t-j) + e(t)$$

$$\text{(ii) } \Delta x(t) = \alpha_0 + \alpha_1 x(t-1) + \sum_j \beta_j \Delta x(t-j) + e(t)$$

$$\text{(iii) } \Delta x(t) = \alpha_0 + \alpha_1 x(t-1) + \sum_j \beta_j \Delta x(t-j) + \delta t + e(t)$$

The corresponding null hypotheses are those of (i)-(iii) of the Dickey-Fuller test presented above.

Step 2. For each case i and the chosen lag order p calculate the test statistic

$$t_i(\alpha_1, p) = [(\alpha_1 - 0)/\text{SE}(\alpha_1)]$$

Step 3. Choose a significance level, α, and critical values, τ_i for each i and p from Fuller (1979) or MacKinnon (1990).

Step 4. Reject the null hypothesis if

$$|t_i(\alpha_1, p)| > |\tau_i|$$

Example

Step 1. For the sake of simplicity only the first regression is estimated with one lag, $p = 1$. The same variable is used as in the previous example.

(i) $\Delta D(t) = -0.0006455D(t - 1) - 0.137865\Delta D(t - 1) + e(t)$

Step 2. The estimated t value is

$$t_1(\alpha_1, 1) = (-0.0006455 - 0)/0.0023221 = -0.2780$$

Step 3. For $\alpha = 0.05$, the critical value for τ_1 is

$$\tau_1 = -1.9421$$

Step 4. Because the critical value of t exceeds the regression value for α, the null hypothesis is accepted with an interpretation similar to that of the DF.

One of the disadvantages of these tests, beyond those already mentioned, is the situation when the number of unit roots present is greater than that stated under the null hypothesis; in this case the actual size of the test or level of significance will be greater than the chosen level of significance, α. To guard against this problem, Dickey and Pantula (1987) suggest a reverse sequential testing procedure. This procedure tests the null hypothesis of m unit roots against the alternative of $m - 1$ unit roots. If the null hypothesis is rejected, then test the null hypothesis of $m - 1$ unit roots. In other words, imagine that you suspect a quadratic trend component in the time series. This implies that there are 2 unit roots, $m = 2$, in the time series. In order to test that assumption, difference the data twice and perform the DF or ADF test. If the null is rejected, then you were incorrect in your assumption and you would have overdifferenced the time series. Just as in the case where one has to be careful of including too few unit roots, one must also guard against including too many.

Extensions have been made to the ADF by Said and Dickey (1985) by including MA terms in the specification. Their method assumes that the number of AR lags, p, and the number of MA lags, q, included in the regression are known. Of course, these numbers usually are not known.

In addition to the above tests, which rely on the parametric nature of $x(t)$, Bhargava (1986) developed a nonparametric method based on von Neumann's ratio, which is a locally very powerful test of stationarity for one-sided alternatives as opposed to the DF and ADF two-sided tests. Phillips and Perron (1986) have developed a nonparametric method test for stationarity that does not rely on the inclusion of lagged dependent variables as in the ADF test. Phillips (1987) has produced a test that is robust against residuals that are heteroscedastic by adjusting the "t ratio," which is then compared to the Dickey and Fuller (1979) critical values. Schwert (1987) suggests that the DF and ADF type tests rely on the crucial assumption that the underlying process is purely autoregressive. It is recognized that although ARMA models have an autoregressive representation, order identification is a major problem. Priestley (1988) employed a test of stationarity using a two-way factor analysis of the variance that enables one to examine the nonconstancy of the mean and variance in one overall testing framework. For further information on the appropriateness of different unit root tests, please refer to Diebold and Nerlove (1989), Cochrane (1991), Dickey and Fuller (1981), Dickey et al. (1986), Evans and Savin (1981), and Said and Dickey (1984).

3. TESTING FOR NORMALITY

The familiar concept of a normal distribution can be combined with independence to provide insights into methods of specifying a time series model. If a time series has been transformed to achieve stationarity and if that series passes the test of normality, then it is possible to specify a linear model to describe the behavior of that time series. The tests provided here prepare the way for this aspect of model specification. Although a number of normality tests exist, only those considered more relevant are presented. For a survey of other available tests, please refer to Conover (1980) and D'Agostino and Stephens (1986).

Normal Distribution Test

One useful property of the normal distribution is that all odd moments greater than two are zero. The measure of skewness equal to the third central moment is zero; however, the fourth central moment, the measure of kurtosis, is around 3. Given this information, various tests for normality can be constructed.

Let $x(t)$ be a random variable with mean μ. Define the rth central sample moment as

$$\mu_r = T^{-1}\sum_t [x(t) - \mu]^r \quad \text{for } r = 1, 2, \ldots, \infty \text{ and } t = 1, \ldots, T \quad (3.1)$$

When $r = 3$, the ratio for the coefficient of skewness becomes $(\beta_1)^{1/2} = \mu_3/(\mu_2)^{3/2}$, and when $r = 4$, the ratio for the coefficient of kurtosis, $\beta_2 = \mu_4/(\mu_2)^2$. As stated above, if the random variable is normal, it is a well-known result that $(\beta_1)^{1/2} = 0$ and that $\beta_2 = 3$.

For a test of skewness, the sample moments of $x(t)$ can be calculated and $(\beta_1)^{1/2}$ can be considered approximately normally distributed with zero mean and standard error, $SE(\beta_1^{1/2}) = (6/T)^{1/2}$. Consequently, the ratio $v_1 = \beta_1^{1/2}/SE(\beta_1^{1/2})$ can be compared with a standard normal variate to test the hypothesis of normality. The null hypothesis is that $x(t)$ is distributed normally. Therefore, the null hypothesis is rejected in the case where v_1 is greater than the selected critical value from a standard normal distribution. Geometrically, negative skewness $(\beta_1^{1/2} < 0)$ is reflected in an extended tail to the left (LS), and positive skewness $(\beta_1^{1/2} > 0)$ is reflected in an extended tail to the right (RS).

A test of normality can also be conducted by using the coefficient of kurtosis, β_2. For large values of T, β_2 is normally distributed, with mean zero and standard error $SE(\beta_2) = (24/T)^{1/2}$. Under the null hypothesis of normality, let $v_2 = (\beta_2 - 3)/SE(\beta_2)$. Reject the null when v_2 is greater than the selected critical value from a standard normal distribution. Kurtosis substantially in excess of 3 indicates heavy tails, an indicator of leptokurtosis.

Jarque-Bera Test

Bera and Jarque (1981) have also utilized these coefficients of skewness and kurtosis to form the following joint statistic

$$S = (T/6)\beta_1 + (T/24)(\beta_2 - 3)^2 \quad (3.2)$$

Under the null hypothesis of normality, S is distributed as a chi-square with 2 degrees of freedom.

Test Procedure

Step 1. The normal distribution test can be performed using the above skewness and kurtosis coefficient for the time series $x(t)$. The coefficients are defined by

$$\text{Skewness} = \frac{T^{-1}\sum_i [x(i) - \mu_x]^3}{T^{-1}\sum_i [x(i) - \mu_x]^{3/2}} = (\beta_1)^{1/2}$$

$$\text{Kurtosis} = \frac{T^{-1}\sum_i [x(i) - \mu_x]^4}{T^{-1}\sum_i [x(i) - \mu_x]^2} = \beta_2$$

Step 2. The following null hypothesis and test statistics are employed. Note that the joint Jarque-Bera test represented by S is just an extension of the normality test.

H_0: $x(t)$ is normal
H_A: $x(t)$ is nonnormal

These tests are based on

(i) $v_1 = (\beta_1)^{1/2}/(6/T)^{1/2}$ Skewness

(ii) $v_2 = (\beta_2 - 3)/(24/T)^{1/2}$ Kurtosis

(iii) $S = (T/6)\beta_1 + (T/24)(\beta_2 - 3)^2$ Jarque-Bera

Step 3. The critical values, τ_i, for the test are obtained from the normal distribution, $N(0,1)$ for tests (i) and (ii). For test (iii) the critical values from the chi-square distribution with 2 degrees of freedom, $\chi^2(2)$, are used.

Step 4. The decision rule employed is to reject the null hypothesis of normality

(i) if $v_i > \tau_i$ for expressions 1 and 2

22

(ii) if $S > \tau_3$ for Jarque-Bera

Example

Step 1. The values of $\beta_1^{1/2}$ and β_2 have been computed for growth rates in party identification $\Delta D(t)$ over the period 1953.2 to 1988.4.

$$(\beta_1)^{1/2} = \frac{(143)^{-1}\sum_i [\Delta D(t) - (-0.0002688)]^3}{(143)^{-1}\sum_i [\Delta D(t) - (-0.0002688)]^{3/2}} = -0.479572$$

$$\beta_2 = \frac{(143)^{-1}\sum_i [\Delta D(t) - (-0.0002688)]^4}{(143)^{-1}\sum_i [\Delta D(t) - (-0.0002688)]^2} = 3.888284$$

Step 2. The computed test statistics are

(i) $v_1 = -0.479572/(6/143)^{1/2} = -2.3412$

(ii) $v_2 = (3.888284 - 3)/(24/143)^{1/2} = 2.1683$

(iii) $S = (143/6)(0.22998) + (143/24)(3.888284 - 3)^2$

$= 5.48119 + 4.7014$

$= 10.18260$

Step 3. The critical values corresponding to the above tests are

(i) $\alpha = 0.05$, $|\tau_1| = 1.96$

(ii) $\alpha = 0.05$, $|\tau_2| = 1.96$

(iii) $\alpha = 0.05$, $\tau_3 = 5.991$

Step 4. Because the computed test values exceed the critical values, all three tests reject normality.

Studentized Range Test

In the case where outliers (i.e., extreme observations) exist in the sample of the distribution to be tested, the kurtosis coefficient can increase quite quickly. The studentized range test can be used to control for this factor. The statistic is based on

$$SR = \frac{\max\{x(i)\} - \min\{x(i)\}}{\{[1/(T-1)]\sum_i [x(i) - \mu_x]^2\}^{1/2}} \tag{3.3}$$

where the range of $x(t)$ is standardized using the standard deviation of $x(t)$. The null hypothesis is that $x(t)$ is normal. Rejection of the null hypothesis occurs when the value of SR exceeds the selected critical values obtained from Pearson and Hartley (1970).

Test Procedure

Step 1. The hypothesis for normality in $x(t)$ is

H_0: $x(t)$ is normal
H_A: $x(t)$ is nonnormal

Step 2. Compute the test statistic

$$SR = \frac{\max\{x(i)\} - \min\{x(i)\}}{\{[1/(T-1)]\sum_i [x(i) - \mu_x]^2\}^{1/2}}$$

Step 3. Obtain the corresponding critical values from Pearson and Hartley (1970).

Step 4. The hypothesis of normality is rejected if SR > τ.

Example

Step 1. The null hypothesis, H_0, implies normality for the sample series $\Delta D(t)$.

Step 2. Compute SR for $\Delta D(t)$ over the period 1953.2 to 1988.4

$$SR = \frac{0.0615317 - (-0.1007477)}{\left\{[1/(143-1)]\sum_i [\Delta D(t) - (-0.0002688)]^2\right\}^{1/2}} = 5.78369$$

Step 3. For the $\alpha = 0.05$, the critical value τ is $\tau = 2.772$.

Step 4. Reject the null hypothesis of normality for $\Delta D(t)$.

In this chapter, we have examined the test procedures for normality. Selection of the appropriate test using the skewness, kurtosis, Jarque-Bera, or Studentized Range test statistic should be based on the presence or absence of outliers in the sample.

4. TESTING FOR INDEPENDENCE

Time series that are independent can be thought of as noncorrelated. From a definitional point of view, if the product of the marginal distributions of two independent random variables is the same as their joint distribution, these two random variables are statistically independent. Our understanding of independence can be coupled with that of normality as part of the time series model specification process mentioned in Chapter 3. Tests for independence can be applied to see if the time series, $x(t)$, displays independence. For example, if the time series $x(t)$ is independent, then the process is completely random, and no deterministic model can be constructed. Tests for normality can also be applied to the stochastic error term, $e(t)$, from the equation for $x(t)$ in order to achieve a better interpretation of model specification.

Consider also the possibility that a time series can appear to be random yet also display regularities (deterministic components). Such behavior has been attributed to time series that are chaotic. For a review of the definitions, conditions, and tests for chaos, please refer to Benhabib and Nishimura (1979), Gleick (1987), or Baumol and Benhabib (1989).

Among the independence tests described in this chapter, the first four are the more traditional: Ljung-Box and Box-Pierce, Turning Points, Runs Analysis, and RV-von Neumann's Ratio. The most recent test is that of Brock, Dechert, and Scheinkman (1986).

Ljung-Box and Box-Pierce Test

One of the older tests for examining independence in a time series is the Portmanteau test that employs the autocorrelation function, defined earlier in Chapter 1. If a time series is independently distributed, then the auto-correlation coefficients $\rho(k)$ are zero for all time lags of the differenced series. An approximate formula for the standard error of $\rho(k)$, denoted by $SE[(\rho k)]$, has been suggested by Kendall and Stuart (1961).

$$SE[\rho(k)] = T^{-\frac{1}{2}} \tag{4.1}$$

Instead of using visual inspection of the sample autocorrelations to check for independence, the Ljung-Box (Ljung & Box, 1978) statistic can be employed. This statistic is defined as

$$Q(k) = T(T + 2)\sum_{m=1}^{k}(T - m)^{-1}\rho^{2}(m) \tag{4.2}$$

where up to k lags of the sample autocorrelation coefficient are present. The null hypothesis implies linear independence in $x(t)$. When the null hypothesis is satisfied, Q is asymptotically $\chi^{2}(k)$ distributed with k degrees of freedom. The null hypothesis of independence is rejected when the value of $Q(k)$ is greater than the selected critical value of the chi-square distribution with k degrees of freedom. One popular modification of the above test known as Box-Pierce (Box & Pierce, 1970), which requires that $T(T + 2)/(T - m)$ in (4.2) be replaced by T, is based on a small sample correction term. Identification of the proper value of k is linked to the sampling frequency of the data set. For example, if the data are annual, it would be expected that a lag of less than 5 years, $k = 5$, would be appropriate for some time series. For monthly data, this lag of 5 years would become one of 60, $k = 60$. Selection of k is, therefore, entirely based on the researcher's a priori knowledge of the "memory" of the process, that is, the correlation between the current period and previous periods.

Test Procedure

Step 1. Select the number of lags k for the estimation of the sample autocorrelation coefficients ρ. Keep in mind that everything relies on the parameter k, and that k is somewhat ambiguous.

Step 2. Given the number of lags, test the null hypothesis, which implies linear independence in $x(t)$

H_0: $\rho(1) = \rho(2) = \ldots = \rho(k) = 0$
H_A: at least one $\rho(k)$ is nonzero

Compute the test statistic, $Q_i(k)$

$$\text{(i) } Q_1(k) = T(T+2)\sum_m (T-k)^{-1}\rho^2(m) \quad \text{Ljung-Box}$$

$$\text{(ii) } Q_2(k) = T\sum_m \rho^2(m) \quad \text{Box-Pierce}$$

Step 3. Choose the significance level α and the critical values of τ, where the latter is distributed as chi-square with k degrees of freedom, $\chi^2(k)$.

Step 4. Reject the null hypothesis of linear independence if

$$Q_i(k) > \tau$$

Example

Step 1. Select $k = 8$ for the party identification variable $\Delta D(t)$ and estimate Q over the sample period of 1953.2 to 1988.4.

Step 2. Compute the test statistics

$$\text{(i) } Q_1(8) = 143(145)\sum_m (143-8)^{-1}\rho^2(m) = 12.87$$

$$\text{(ii) } Q_2(8) = 143\sum_m \rho^2(m) = 12.21$$

Step 3. For $\alpha = 0.05$, the critical value obtained from a chi-square distribution with 8 degrees of freedom is

$$\chi^2(8) = 15.5073$$

Step 4. In this case, accept the null of independence because $Q_1(8)$ and $Q_2(8)$ are greater than the critical value, τ .

Besides the problem of lag selection, this test examines the presence of serial correlation only under the alternative. This depends on the assumption that the data are distributed normally and are stationary. If one or both of these two assumptions are incorrect, then the power of the test decreases.

Turning Point Test

In order to determine if a sequence of observations is independent, we can examine the number of turning points in the sequence. For the sequence, $x(1), \ldots, x(T)$, a turning point at time t, $1 < t < T$, occurs when $x(t - 1) < x(t)$ and $x(t) > x(t + 1)$ or $x(t - 1) > x(t)$ and $x(t) < x(t + 1)$. Let V be the number of turning points in the sequence $x(t)$. If $x(t)$ is independent, then the probability of a turning point at time t is $2/3$. In order to test the null hypothesis that $x(t)$ is independent or random, the following test procedure must be followed.

Test Procedure

Step 1. Count the actual number of turning points in the stationary process $x(t)$. Based on the following definitions compute the mean and variance according to

$$E\{V\} = 2(T - 2)/3$$

$$\sigma_V^2 = (16T - 29)/90$$

Step 2. Test the following null hypothesis with the test statistic Z

H_0: $x(t)$ is independent
H_A: $x(t)$ is dependent

The test statistic is

$$Z = |V - E\{V\}|/\sigma_V$$

Step 3. For a chosen significance level, α, the critical value τ is based on the standard normal distribution, $N(0,1)$.

Step 4. Reject the null hypothesis if $Z > |\tau|$.

Example

Step 1. For the variable $\Delta D(t)$ over the period 1953.3 to 1988.4, the actual number of turning points, mean, and variance are computed as follows

$$\text{Count} = 96$$

$$E\{V\} = 2(142 - 2)/3 = 93.33$$

$$\sigma_V^2 = [16(142) - 29]/90 = 24.92$$

Step 2. The test statistic is

$$Z = |96 - 93.33|/4.992 = 0.5348$$

Step 3. For the chosen significance level, $\alpha = 0.05$, the critical value is

$$|\tau| = 1.96$$

Step 4. Because $|Z| < |\tau|$, accept the null hypothesis that $\Delta D(t)$ is independent.

Runs Test

This test searches for independence by examining the frequency of certain repetitive patterns with repeated trials. If a sample from such a process is random, then too many or too few runs are unlikely. The meaning of too few or too many runs can be illustrated in the following example. Consider the variable, $x(t)$, which consists of 12 observations of both positive and negative values. Replace each observation with the sign of that observation. For example,

$$(+)(+)(+)(-)(+)(-)(-)(-)(-)(+)(-)(+)$$

In order to determine the actual number of runs, R, we count the number of changes from $(+)$ to $(-)$ and from $(-)$ to $(+)$. In this example, there are 7 runs: 4 runs of positive values with the number of elements of these runs 3,1,1,1; and 3 runs of negative values with the number of elements of these runs 1,4,1. Under the null hypothesis of independence the expected number of runs can be estimated as

$$m = [T(T + 1) - \sum_i n^2(i)]/T \quad i = 1, 2, 3 \qquad (4.3)$$

where T is the total number of observations, the $n(i)$ are the number of sign changes of each type, and $i = 1, 2, 3$ represents the types of change positive, negative and zero. The variance of m is given by

$$\sigma^2(m) = \frac{\sum_i n^2(i)\left[\sum_i n^2(i) + T(T + 1)\right] - 2T\sum_i n^3(i) - T^3}{T^2(T - 1)} \qquad (4.4)$$

For large T, the sampling distribution of m is approximately normal. A standardization can be employed in order to utilize the standard normal distribution by calculating

$$Z = [(R + 0.5) - m]/\sigma(m) \qquad (4.5)$$

where R is the actual number of runs. Thus, one could compare the value of Z with the critical values of the standard normal, $N(0,1)$.

Test Procedure

Step 1. Compute the expected number of runs m as well as the variance of m for the process $x(t)$.

$$m = [T(T + 1) - \sum_i n_i^2]/T$$

where n_i is defined as the number of positive(+), negative(−) or zero (0) sign changes. The expected variance is

$$\sigma_m^2 = \frac{\sum_i n_i^2\left[\sum_i n_i^2 + T(T + 1)\right] - 2T\sum_i n_i^3 - T^3}{T^2(T - 1)}$$

Step 2. The null hypothesis concerning $x(t)$ reflects independence

H_0: $x(t)$ is independent
H_A: $x(t)$ is dependent

The test of this hypothesis is based on the given Z statistic where R is the actual number of runs.

$$Z = [(R + 0.5) - m]/\sigma_m$$

Step 3. For a given significance level, α, the critical value, τ, to be compared with Z is obtained from the standard normal distribution, $N(0,1)$.

Step 4. The decision rule is to reject the null hypothesis of independence if $|Z| > |\tau|$.

Example

Step 1. The expected number of runs and variance was calculated for the variable $\Delta D(t)$ over the period 1954.1 to 1988.4.

$$m = \{136(136 + 1) - 890]/136 = 71.5589$$

$$\sigma_m^2 = \frac{890[890 + 136(136 + 1)] - 2(136)(603142) - (136)^3}{136^2(136 - 1)} = 31.4241$$

Step 2. The test statistic for testing the null hypothesis of independence, with $R = 77$, the actual number of runs in $\Delta D(t)$, is

$$Z = [(77 + 0.5) - 71.5589]/5.6057 = 1.05983$$

Step 3. For the significance level chosen, $\alpha = 0.05$, the critical value from a standard normal distribution, $N(0,1)$, is

$$|\tau| = 1.96$$

Step 4. Because $|Z| < |\tau|$, the null hypothesis of independence is accepted.

Rank Version of the von Neumann Ratio Test

An alternative test method to the parametric methods above, the rank von Neumann ratio test, evaluates the randomness of a differenced variable, $\Delta x(t)$. The rank version of the von Neumann ratio test procedure is as follows. Let $R(i)$ be the rank associated with the ith observation in a sequence of T observations and define the rank version of the von Neumann ratio (RVN) to be

$$\text{RVN} = \frac{\sum_i [R(i) - R(i+1)]^2}{\sum_j [R(j) - \mu_R]^2} \qquad (4.6)$$

where $i = 1$ to $T - 1$, $j = 1$ to T and μ_R is the mean of $\Delta x(t)$. Bartels (1982) provides a relatively straightforward explanation according to which runs tests should be less powerful than a test based on ranks, because runs tests completely ignore the magnitudes of the observations. Critical values for this test are also provided in Bartels (1982).

Test Procedure

Step 1. Rank the values of $\Delta x(t)$ in ascending order, and call this new variable $R(i)$.

Step 2. Similar to the runs test, the null hypothesis stipulates that $x(t)$ is independent

H_0: $\Delta x(t)$ is independent
H_A: $\Delta x(t)$ is dependent

The test of this hypothesis is based on the RVN test statistic

$$\text{RVN} = \frac{\sum_i [R(i) - R(i+1)]^2}{\sum_i [R(i) - \mu_R]^2}$$

Step 3. The critical value, τ, to be compared to RVN is found in Bartels (1982).

Step 4. Reject the null hypothesis that $x(t)$ is independent if

$$RVN > \tau.$$

Example

Step 1. The observations for $\Delta D(t)$ was ranked in ascending order over the period from 1953.3 to 1988.4.

Step 2. The test statistic is

$$RVN = \frac{\sum_i [R(i) - R(i+1)]^2}{\sum_j [R(j) - (-0.00419)]^2} = 0.714508$$

Step 3. For the significance level of $\alpha = 0.05$, the critical value is obtained from Bartels (1982).

$$\tau = 1.67$$

Step 4. Because RVN < τ, the null hypothesis that $\Delta D(t)$ is independent is accepted.

Brock, Dechert, and Scheinkman (BDS) Test

The literature on testing for chaos contains the Brock, Dechert, and Scheinkman (1986) statistic (BDS), which tests the null hypothesis of independence for a time series $x(t)$ by utilizing the concept of *spatial correlation*. To examine this "spatial" correlation, the time series, $x(t)$, must be *embedded* in *m-space* by constructing the following vector,

$$x^m(t) = [x(t), \ldots, x(t - m + 1)] \quad t = 1, 2, \ldots, T - m + 1 \quad (4.7)$$

Takens (1980) suggested this concept of embedding, which can be illustrated for the data set $x(t)$ where $t = 1, 2, \ldots, 5$ and $m = 3$.

$$x^3(1) = [x(1), x(2), x(3)]$$

$$x^3(2) = [x(2), x(3), x(4)]$$

$$x^3(3) = [x(3), x(4), x(5)]$$

In this case the "embedding" operation creates three new three-dimensional vectors $x^3(1)$, $x^3(2)$, and $x^3(3)$. Because it is required that the vectors all be of equal length with this method, $m - 1$ data points are lost as in the above illustration. Thus, embedding enables one to take a univariate variable and "embed" it into a higher dimensional space. The choice of m for the dimensionality of the vectors is subjective. The motivation behind this is to examine correlation in a spatial context.

The dependence of $x(t)$ is examined through the concept of the correlation integral, a measure that examines the distances between points in our three-dimensional example. For each embedding dimension, m, and choice of epsilon, ε, the correlation integral is defined by

$$C(\varepsilon, m, T) = [T_m(T_m - 1)]^{-1} \sum_{t \neq s} I[x^m(t), x^m(s); \varepsilon] \qquad (4.8)$$

where $T_m = T - m + 1$, t and s both range from 1 to $T - m + 1$ in the summation and are restricted such that $t \neq s$. The indicator function in (4.8) is

$$I[x^m(t), x^m(s); \varepsilon] = 1 \text{ if } \|x^m(t) - x^m(s)\| \leq \varepsilon \qquad (4.9)$$

$$= 0 \text{ otherwise}$$

The term in absolute bars is called a *metric* or norm where the maximum norm is given by $\|x\| = \max_{0 \leq i \leq m-1} |x_i|$. Thus, the correlation integral will measure the fraction of total pairs of $[x^m(t), x^m(s)]$ for which the distance between $x^m(t)$ and $x^m(s)$ is no more than epsilon, ε. For example, the distance between $x^3(1) = [x(1), x(2), x(3)]$ and $x^3(3) = [x(3), x(4), x(5)]$ is computed by $x^3(1) - x^3(3) = [x(1) - x(3), x(2) - x(4), x(3) - x(5)]$. One then chooses the maximum element of this resulting vector. This maximum element is then compared to the chosen value of ε.

If this maximum $> \varepsilon$, then the pair is counted; otherwise it is not. As we vary t and s, we count the number of pairs that satisfy the condition. We then divide this total number of pairs that satisfy the condition by the total number of possible pairs making sure not to count any pairs twice. This expression then yields the fraction of pairs that are within distance ε of each other. If the value of ε is chosen such that all pairs satisfy the condition then $C(m, T, \varepsilon) = 1$. Of course, if ε is chosen such that the condition is never satisfied, then $C(m, T, \varepsilon) = 0$. This is why the correlation integral is interpreted as a measure of spatial correlation.

Brock and Baek (1991) have demonstrated that under the null hypothesis of independence, the BDS test statistic will be asymptotically standard normally distributed

$$W(\varepsilon, m, T) = T^{\frac{1}{2}}[C(\varepsilon, m, T) - C(\varepsilon, 1, T)^m]/V^{\frac{1}{2}} \qquad (4.10)$$

where the variance is defined by

$$V = 4\{K(\varepsilon)^m \qquad (4.11)$$

$$+ 2[\sum_i K(\varepsilon)^{m-i}C(\varepsilon)^{2i} + (m-1)^2 C(\varepsilon)^{2m} - m^2 K(\varepsilon)C(\varepsilon)^{2m-2}]\}$$

for $i = 1$ to $m - 1$

Here the indicator functions $C(\varepsilon) = E\{I[x(i), x(j); \varepsilon]\}$ and $K(\varepsilon) = E\{I[x(i), x(j); \varepsilon]I[x(j), x(k); \varepsilon]\}$ are defined similarly.

Notice that the statistic is a function of two unknowns, the embedding dimension m and the epsilon, ε. An important relation exists between the choice of ε and m, and the small sample properties of the BDS statistic. For a given m, ε cannot be too small because $C(\varepsilon, T)$ will capture too few points; similarly ε cannot be too large, to prevent $C(\varepsilon, T)$ from capturing too many points. Often in empirical practice, ε is set in terms of the standard deviation of the data, with $\varepsilon = 1$ implying that it is one standard deviation of the data. Some researchers transform the data, $x(t)$, to the unit interval, [0,1], and set epsilon, $\varepsilon = 0.9^i$, where i is from $1, \ldots, m - 1$. The embedding dimension, m, typically is chosen over the range 1 to 15. Thus, the BDS test is computed over a grid of epsilons and embedding dimensions. Because many tests are being performed that may give contradictory information, care must be undertaken in interpretation.

To insure that the BDS test is not merely reflecting linear dependence in the data, one can pre-filter the data. Brock (1987) shows that the asymptotic distribution of the BDS statistic applies to residuals of linear regressions as well as to the original data. Thus any linear dependence that exists can be removed before applying the test to the data. This itself amounts to a diagnostic on the residuals, useful for testing the stationary and independent assumptions discussed earlier.

Test Procedure

Step 1. Transform $x(t)$ to the unit interval, $[0,1]$.

Step 2. Choose a grid of values for $\varepsilon_i = 0.9^i$ and $m = i + 1$.

Step 3. Again the null hypothesis reflects independence

H_0: $x(t)$ is independent
H_A: $x(t)$ is dependent

The test statistic W provides the basis for the hypothesis test

$$W(\varepsilon, m, T) = \frac{(T - m + 1)^{\frac{1}{2}}[C(\varepsilon, m, T) - C(\varepsilon, m, T)^m]}{V^{\frac{1}{2}}}$$

Step 4. For the given significance level, α, the critical values to test H_0 are based on the number of observations dividing the selected embedding dimension.

(i) If $T/m > 200$, use the standard normal distribution, $N(0,1)$, for the critical value, τ.
(ii) If $T/m < 200$, use τ from tables in Brock, Hsieh, and Lebarron (1991).

Step 5. Reject the null hypothesis of independence if $|W| > |\tau|$

Example

Step 1. Transform the data for the sample variable $\Delta D(t)$ over the period 1953.2 to 1988.4 according to

$$u\Delta D(t) = \frac{\Delta D(t) - (-0.1007477)}{0.0615317 - (-0.1007477)}$$

Step 2. Select the following values for the distance metric, ε, and the embedding dimension, *m*,

$$\varepsilon = 0.9 \text{ and } m = 2$$

Step 3. The null hypothesis of independence is tested from

$$W = \frac{(143 - 2 + 1)^{1/2}[C(0.9, 2, 143) - C(0.9, 1, 143)^2]}{0.00005043} = -4.91307$$

Step 4. Because of the incompatibility of our data and the assumption inherent in the critical values from Brock et al. (1991), the standard normal was used for illustrative purposes.

$$\alpha = 0.05, |\tau| = 1.96$$

Step 5. Because the value of |*W*| is greater then |τ|, the null hypothesis of independence is rejected.

Brock et al. (1991) state that their Monte Carlo simulations demonstrate that the standard normal distribution is a good approximation for samples of more than 500 observations. When the data are normally distributed and are fairly robust to problems of skewness and leptokurtosis, empirical practice often dictates the use of a smaller number of observations. In this case as well as when an ARCH or a GARCH effect are present in the residuals, the asymptotic distribution of the BDS test may not be normal. As was reported in Brock et al. (1991), if the critical values from the standard normal are to be used, the number of observations should be greater than 500; the dimension should be $m \leq 5$; and epsilon should be between 0.5 and 2 standard deviations of the data.

The reader should note that only the BDS test rejected the null hypothesis of independence in the examples. In contrast the Box-Pierce, turning point, runs analysis, and rank VNR accept it. Because the Box-Pierce, turning point, runs, and rank VNR test primarily for linear correlation in the presence of nonnormality, they may accept the null hypothesis of independence even in the presence of nonlinear dependence.

5. TESTING FOR LINEAR
OR NONLINEAR DEPENDENCE

If a time series lacks independence, it possesses some form of dependence, either simple or complex, weak or strong. In the simplest case this dependence may reflect a linear trend; more complex processes may require a nonlinear representation. Even the latter can be simple, such as an exponential. In the past, time series dependence was often evaluated in the frequency domain using tests of spectral analysis. More recently, this approach has been updated, for example by Subba Rao and Gabr (1980, 1984) and by Hinich (1982) who emphasize a bispectrum test based on bispectral analysis. In this study, time series dependence has been examined mostly using tests in the time domain. Most of these dependence tests are parametric and are based on what has become known as the Volterra expansion. These tests evaluate linear as compared to nonlinear dependence in a time series; an exception is the Hsieh Test, which additionally identifies the actual type of nonlinear dependence. Nonparametric tests such as the neural network test (e.g., Lee et al., 1993) can also be performed but are not included here.

The premise of the following tests involves the specification of a linear model under the null hypothesis that is then estimated and the residuals computed. Regression techniques are then used to examine the correlation between the residuals and power transformations of the variable in question. This method also enables the BDS test from the previous chapter to be included in tests for nonlinear dependence. For example, if a linear model is the correct specification, then the residuals will be independent (e.g., BDS test). As in the previous chapter, not all possible tests of dependence are included here, only the more popular ones. The reader is referred to Chan and Tong (1986), deGooijer and Kumar (1992), and Granger (1991) for a complete and detailed treatment.

Keenan Test

The Keenan (1985) test for nonlinear dependence stems from work by Priestley (1980) on a Volterra series expansion in the moving-average component that allows a mean zero process $x(t)$ to be represented as follows

$$x(t) = \sum_i \beta_i e(t) + \sum_i \sum_j \delta_{ij} e(t-i) e(t-j) \qquad (5.1)$$

$$+ \sum_i \sum_j \sum_k \theta_{ijk} e(t-i) e(t-j) e(t-k) + \ldots$$

Here, $x(t)$ is nonlinear if any of the higher order coefficients are nonzero (i.e., high order with respect to the interaction or multiplicative terms in (5.1)); otherwise, the representation is linear. Keenan's test, thus, focuses on the presence of these multiplicative terms in (5.1) under the alternative hypothesis.

The actual form of this test requires that one distinguish between linearity versus a second-order Volterra expansion, by examining $\theta_{ijk} = 0$ as well as the coefficients on higher orders. That is, the null hypothesis of linearity for a mean zero stationary process $x(t)$ is compared to the alternative hypothesis reflecting the expansion. More formally, the null and alternative hypotheses can be stated as

H_0: $x(t)$ has an ARMA(p,q) representation
H_A: $x(t) = \sum_i a_i e(t-i) + \sum_i \sum_j b_{ij} e(t-i) e(t-j)$

The theory of autoregressive moving-average models [ARMA(p,q)] is explained in the next chapter. For the moment assume that the process $x(t)$ is linear, and an ARMA(1,1) is an "adequate" fit to the time series. If the time series is indeed nonlinear, then upon fitting a linear model to this time series the hypothesized nonlinearity would be "swept" into the residuals of the ARMA(1,1) specification. Therefore, if we examine the residuals by specifying a nonlinear specification under the alternative hypothesis, it may be possible to detect this nonlinearity. In order to accomplish this the following testing procedure is proposed.

Test Procedure

Step 1. Fit the model under the null hypothesis and calculate the fitted values $x(t)$, the residuals, $e(t)$, and the residual sum of squares $SSE_e = \sum e^2(t)$.

H_0: $x(t)$ is an ARMA(p,q) process
H_A: $x(t)$ is not an ARMA(p,q) process

Step 2. From these fitted values, perform the regression of $x^2(t)$

$$x^2(t) = \alpha_0 + \sum_i \alpha_i x(t - i) + v(t) \quad \text{for } i = 1, \ldots, m$$

It has been suggested that the selection of m is not an important issue.

Step 3. Calculate the residuals, $v(t)$, and the residual sum of squares, SEE_v .

Step 4. Perform the following regression to obtain the value of β. The residuals under the null hypothesis become

$$e(t) = \beta v(t) + w(t)$$

Step 5. The residuals $e(t)$ (i.e., residuals under the null) and $v(t)$ are reflected in the null and alternative hypothesis of Step (1). Test the null hypothesis according to the statistic

$$F = \frac{\beta^2(SSE_v)(T - 2m - 2)}{(SSE_e) - (\beta^2 SSE_v)}$$

Step 6. For a given significance level, α, obtain the critical value to compare to F from the table of the F distribution with 1 and $T - 2m - 2$ degrees of freedom, $F(1, T - 2m - 2)$.

Step 7. Reject the null hypothesis and assume nonlinear dependence if $F > \tau$.

Example

Step 1. Over the period, 1953.3 to 1988.4 an AR(1) model with no constant was fitted to the party identification variable $\Delta D(t)$. From this model the resulting fitted values for $\Delta D(t)$, the residuals, $e(t)$, and the residual sum of squares SSE were calculated,

$$SSE_e = 0.110051$$

Step 2. For $m = 1$, a regression was then performed as stated in step (2) above:

$$x^2(t) = \alpha_0 + \alpha_1 x(t - 1) + v(t)$$

Step 3. The residuals from Step 2 were calculated along with the residual sum of squares, $SEE_v = 0.000000057$.

Step 4. The following regression was performed in order to obtain β for the test statistic

$$e(t) = -79.0399v(t) + w(t)$$

Step 5. The test statistic is then calculated in order to test the null hypothesis

$$F = \frac{(-79.0399)^2(0.000000057)[142 - 2(1) - 2]}{(0.110051) - (-79.0399)^2(0.000000057)} = 4.4508$$

Step 6. For the chosen significance level, $\alpha = 0.05$, the critical value is from $F[1, 142 - 2(1) - 2]$. By interpolation, the critical value is

$$\tau = 3.91$$

Step 7. Because the value of $F > \tau$, the null hypothesis that $\Delta D(t)$ is linear is rejected.

The advantages of this test are that it is quick and easy to implement, because only the lag length, m, has to be selected, and its results are robust. The main disadvantage is that it needs a truncated functional expansion estimated under the alternative hypothesis and this brings into question estimation issues and potential specification error.

The RESET test is another version of the Ramsey (1969) test, upon which the Keenan test is based. It examines the correlation between $x^k(t)$ and $x(t)$ in Step 2 of the Keenan test procedure. The k term is a power transformation on $x(t)$ greater than 2. Lee, White, and Granger (1993) discuss a Lagrangian multiplier version of this test.

Luukkonen Test

Variants of the Keenan test have been proposed by Luukkonen, Saikkonen, and Terasvirta (1988). For example, one can replicate the following process as in the case of Keenan with the null hypothesis that $x(t)$ has an AR(p) representation. Instead of an intermediate regression being performed as in Step 2 of the Keenan test, these tests regress the residuals from the null hypothesis on the nonlinear specification under the alternative hypothesis. This process is similar to fitting (5.1) as a Volterra expansion in $x(t)$ alone.

Test Procedure

Step 1. Perform a regression under the null hypothesis

$$x(t) = \alpha_0 + \sum_i \alpha_i x(t-i) + e(t) \quad \text{for } i = 1, \ldots, p$$

H_0: $x(t)$ is an AR(p) process
H_A: $x(t)$ is not an AR(p) process

Step 2. Based on the above regression, calculate the residuals, $e(t)$ and the residual sum of squares, SSE_e.

Step 3. Perform one of the following regressions

(i) $e(t) = \alpha_0 + \sum_j \alpha_j x(t-j) + \sum_i \sum_j \beta_{ij} x(t-i)x(t-j) + v_1(t)$

(ii) $e(t) = \alpha_0 + \sum_j \alpha_j x(t-j) + \sum_i \sum_j \beta_{ij} x(t-i)x(t-j)$

$$+ \sum_i \sum_j \delta_{ij} x(t-i)x^k(t-j) + v_2(t)$$

for $i, j = 1, \ldots, p$ and $k = 2, 3$

$$(\text{iii}) \ e(t) = \alpha_0 + \sum_j \alpha_j x(t-j) + \sum_i \sum_j \beta_{ij} x(t-i)x(t-j)$$

$$+ \sum_j \delta_j x^3(t-j) + v_3(t)$$

Step 4. Calculate the test statistic and test the above hypothesis

$$S_i = (T-p)[1 - (SSE_{vi}/SSE_e)] \quad \text{for } i = 1, 2, 3$$

where S_1 is known as the Tsay (1986) test statistic.

Step 5. Critical values, τ_i for S_i depend on which regression has been selected.

If (i) is chosen in Step 3, then τ_1 is $\chi^2[p(p+1)/2]$
If (ii) is chosen in Step 3, then τ_2 is $\chi^2[p^2 + p(p+1)/2]$
If (iii) is chosen in Step 3, then τ_3 is $\chi^2[p + p(p+1)/2]$

Step 6. Reject the null hypothesis of linear dependence if $S_i > \tau$.

Example

Step 1. Perform the regression of a simple autoregressive model for $\Delta D(t)$ with one lag over the sample period 1953.3 to 1988.4. This model defines the null hypothesis

$$\Delta D(t) = \alpha_1 \Delta D(t-1) + e(t)$$

Step 2. Calculate the residuals $e(t)$ and the sum of squares, SSE_e

$$SSE_e = 0.110051$$

Step 3. Perform the three suggested regressions

$$(\text{i}) \ e(t) = \alpha_1 \Delta D(t-1) + \beta_{11} \Delta D^2(t-1) + v_1(t)$$

$$(\text{ii}) \ e(t) = \alpha_1 \Delta D(t-1) + \beta_{11} \Delta D^2(t-1) + \delta_{11} \Delta D^2(t-1)\Delta D(t-1) + v_2(t)$$

(iii) $e(t) = \alpha_1 \Delta D(t - 1) + \beta_{11} \Delta D^2(t - 1) + \delta_1 \Delta D^3(t - 1) + v_3(t)$

Step 4. Compute the test statistic for each regression

$$S_1 = (142)[1 - (0.109678/0.110051)] = 0.4812$$

$$S_2 = (142)[1 - (0.107584/0.110051)] = 3.1832$$

$$S_3 = (142)[1 - (0.050874/0.110051)] = 76.4005$$

Step 5. For the significance level, $\alpha = 0.05$, the critical values for S_i are

$$\tau_1 \sim \chi^2(1) = 3.8416$$

$$\tau_2 \sim \chi^2(2) = 5.9914$$

$$\tau_3 \sim \chi^2(2) = 5.9914$$

Step 6. Reject the null for S_3 that $\Delta D(t)$ is an AR(1) process but accept the null hypothesis for S_1 and S_2.

The main advantage of this LM test is that it can be used with either a known or unknown delay parameter d. (For a discussion on the delay parameter and threshold models consult Chapter 7). The disadvantages are that the researcher must fix p, and that the size of the regressions in step (3) can grow quite large very rapidly, that is, increases in the number of interaction terms. As an alternative to the specification of the truncated Volterra expansions, Petruccelli and Davies (1986) have proposed a test based on the specifications of threshold models introduced in Chapter 7. Their method relies on the estimation of cumulative sums (CUSUM). Although it is a desirable test to perform in addition to ones mentioned above, it suffers from a lack of critical values from some parametric distribution, that is, normal, chi-square, and so forth. Instead, the critical values are computed based on one's own Monte Carlo simulations.

McLeod-Li Test

The McLeod and Li (1983) Portmanteau test for nonlinear dependence is conducted by examining the Box-Pierce Q statistic of the squared residuals from an ARMA representation. Instead of using the residuals from a linear representation, the raw data can be examined through the use of the k autocorrelation coefficients for $\{x(t)\}$, $\{|x(t)|\}$ and $\{x^2(t)\}$. The power of this test is, of course, a function of normality. That is, the power of the test is good in the presence of normality for $x(t)$. The Q statistic for each of these three transformed data series can be used to examine the presence of serial correlation. For example, Granger and Newbold (1986) have suggested that if

$$\rho_x(k) = [\rho_x(k)]^2 \quad \text{for all } k$$

then the time series $x(t)$ is linear. Alternatively, non-equality implies that $x(t)$ is nonlinear. Thus the sample autocorrelation function can provide evidence for or against the linearity hypothesis.

Test Procedure

Step 1. Select lag length k based on the sample frequency and estimate the autocorrelation function of $x^2(t)$.

$$\rho_x^x(k) = \frac{\sum_t [x^2(t) - \sigma^2][x^2(t+k) - \sigma^2]}{\sum_t [x^2(t) - \sigma^2]^2}$$

where σ^2 is the variance of $x^2(t)$.

Step 2. The null hypothesis now reflects independence

H_0: $x^2(t)$ is independent
H_A: $x^2(t)$ is dependent

The same hypothesis and testing procedure can be followed for $|x(t)|$. Compute the following statistic to test the hypothesis,

$$Q_{xx}(k) = T(T+2)\sum_m (T-m)^{-1}\rho_{xx}^2(m)$$

Step 3. For a given significance level, α, obtain the critical value τ to compare to $Q_{xx}(k)$ from the table of the chi-square distribution, $\chi^2(k)$.

Step 4. Reject the null hypothesis of linear dependence if $Q_{xx}(k) > \tau$.

Example

Step 1. Estimate the autocorrelation function for the example variable $\Delta D(t)$ over the period 1953.2 to 1988.4. A lag value 8 is chosen to represent 2 years of quarterly data.

Step 2. The test statistic for the null hypothesis is

$$Q_{\Delta D}(8) = (143)(143 + 2)\sum_{m}(143 - m)^{-1}\rho_{\Delta D}^2(m) = 4.97$$

Step 3. The critical value of τ obtained from $\chi^2(8)$ is $\tau = 15.5073$.

Step 4. Because $Q_{\Delta D}(8) < \tau$, accept the null hypothesis of independence for $\Delta D^2(t)$.

Hsieh Test

Once it is established that some type of nonlinearity exists, the Hsieh (1989) test discriminates between two different types of nonlinearity: additive and multiplicative. Multiplicative dependence implies that the conditional expectation of the residuals given past lags of the variable, $x(t)$ and the residuals, $e(t)$, is zero,

$$E[e(t)|x(t - 1), \ldots, x(t - k), e(t - 1), \ldots, e(t - k)] = 0 \qquad (5.2)$$

Additive dependence implies that the same conditional expectation is nonzero

$$E[e(t)|x(t - 1), \ldots, x(t - k), e(t - 1), \ldots, e(t - k)] \neq 0 \qquad (5.3)$$

Additive dependence occurs when the nonlinearity enters only through the mean of the process. Examples of such processes include the nonlinear moving-average model, the threshold and exponential autoregressive models, and the bilinear model. These will be explained in Chapter 7. Multiplicative dependence occurs when the source of the nonlinearity is in the variance of the process. An example of this would be the autoregressive conditional heteroscedastic (ARCH) models and the general form of the autoregressive conditional heteroscedastic (GARCH) models.

Both additive and multiplicative nonlinearity imply that the squared residuals, $e^2(t)$, from a linear autoregression are correlated with their own lags. The test requires that we set up multiplicative nonlinearity as the null hypothesis, which implies that the third-order correlation coefficient, $\rho_{eee}(i, j) = 0$ for all $i, j > 0$ and then we test it against the composite alternative hypothesis that $\rho_{eee}(i, j) \neq 0$ for some $i, j > 0$. The correlation is based on the residuals from a linear specification, for example, an AR(p) model. Once again the test is based on the fact that by fitting a linear model to the data, the inherent nonlinearity has been swept into the residuals.

The null hypothesis appears in this form because this test is designed to be rejected only in the presence of additive nonlinearity but not of multiplicative nonlinearity. In contrast, the Tsay (1986) test, where $S(1)$ is used in case (ii) of the Luukkonen test, is designed to detect any type of nonlinearity, whether it is additive or multiplicative.

The Hsieh test has also been used to evaluate the hypothesis as to whether the third moment (i.e., skewness measure) is different from zero, up to the fourth lag. The null hypothesis of multiplicative nonlinearity is rejected if the value of (5.4) is greater than the selected critical value of the standard normal distribution. If the variable $x(t)$ is not linear dependent, but not independent, the "third moment" test can be written as

$$V(i, j) = \frac{T^{1/2}\rho_{xxx}(i, j)}{w(i, j)} \tag{5.4}$$

where

$$\rho_{xxx}(i, j) = \frac{(1/T)\sum x(t)x(t-i)x(t-j)}{\left[(1/T)\sum x^2(t)\right]^{1.5}}$$

and

$$w(i, j) = \frac{(1/T)\sum x^2(t)x^2(t-i)x^2(t-j)}{\left[(1/T)\sum x^2(t)\right]^3}$$

Critical values for (5.4) are based on the standard normal distribution. As was previously stated, this third moment test can also be performed on the residuals from a linear regression. In this case, (5.4) can be rewritten with the residuals $e(t)$ in the place of $x(t)$. Because of the results of the previous two chapters, we use the former approach in our example.

Test Procedure

Step 1. Fit an AR(p) model to the data to remove the linearity. Estimate the third-order moment of the residuals $e(t)$

$$\rho_{eee}(i,j) = \frac{(1/T)\sum_t e(t)e(t-i)e(t-j)}{\left[(1/T)\sum e^2(t)\right]^{1.5}}$$

If a linear specification does not exist for $x(t)$, the residuals $e(t)$ in the expressions can be replaced by the variable $x(t)$.

Step 2. Formulate the following null and alternative hypotheses

H_0: $x(t)$ possess a multiplicative or variance nonlinearity
H_A: $x(t)$ possess an additive or mean nonlinearity

Step 3. In order to test the null hypothesis, compute the following test statistic

$$V(i,j) = \frac{T^{1/2}\rho_{eee}(i,j)}{w(i,j)}$$

where $w(i,j)$ can be consistently estimated by

$$w(i,j) = \frac{(1/T)\sum e^2(t)e^2(t-i)e^2(t-j)}{\left[(1/T)\sum e^2(t)\right]^{3}}$$

Step 4. For a selected significance level α, find the critical value τ for testing the null hypothesis of multiplicative nonlinearity using the standard normal distribution, $N(0,1)$.

Step 5. Reject the null hypothesis, if $|V(i, j)| > |\tau|$

Example

Step 1. Over the period of 1952.1 to 1988.4, the third moment of the growth rate in the party identification variable $\Delta D(t)$ was estimated by its sample moments.

$$\rho_{\Delta D(t)}(1, 1) = \frac{(1/142)\sum_t \Delta D(t)\Delta D(t-1)\Delta D(t-1)}{\left[(1/142)\sum \Delta D^2(t)\right]^{1.5}}$$

Step 2. The null hypothesis is that $\Delta D(t)$ possesses a nonlinearity in the variance of the process.

Step 3. The computed test statistic is

$$V(1, 1) = \frac{(142)^{1/2}(-0.0448)}{(7.5210)} = -0.0641$$

Step 4. For the chosen significance level, $\alpha = 0.05$, the critical value is

$$|\tau| = 1.96$$

Step 5. Because $|V(1,1)| < 1.96$, accept the null hypothesis of multiplicative nonlinearity.

Hsieh's third moment test has the disadvantage that several lags have to be tested and that the selection of lags i and j for the test statistic is ambiguous. Once again a grid of test statistics is evaluated and one looks to the majority of the test results to support an outcome.

In conclusion, one recent study shows that the criteria for choosing among these tests are not particularly clear. Lee et al. (1993) provide empirical comparisons between these tests and several others, including the Keenan, Tsay, RESET, White, McLeod-Li, BDS, Bispectrum, and neural network tests. Interested readers should consult their findings.

6. LINEAR MODEL SPECIFICATION

Univariate linear time series models (ULT) represent the simplest case where the relations between $x(t)$ and past values of itself can be represented as a linear equation. Models of this type can explain or forecast $x(t)$ in the form of linear trend, cyclical, or seasonal models. However, as stated, our interest is with linear models that reduce the explanation of $x(t)$, as suggested by Priestley (1981), to a process of the following form

$$f[x(t \pm 1), x(t \pm 2), \ldots, x(t \pm p)] = e(t) \qquad (6.1)$$

This equation requires the consideration of the irregular component in the form of $e(t)$, which has been defined as an independent, stationary process. In this chapter the identification and resulting estimation of these models has been simplified as much as possible. More detailed explanations of their construction already exist, for example, in works by Granger and Newbold (1986), Labys and Granger (1970), McLeary and Hay (1980), or Mills (1990).

To specify linear time series models, we employ the various tests featured in the steps of the model construction procedure that have been discussed in the previous chapters. To begin with, the time series, $x(t)$, employed must be stationary in the wide sense in order to interpret correctly the autocorrelation function (ACF). Because the specification of the function defined in (6.1) is assumed linear in this chapter, the most important test to be performed in this regard is determining how many lags [i.e., the number of lags p as in (6.1) or the value of k in the autocorrelation function (1.4)] to be included in the specification. This is often referred to as the order of the models.

Autoregressive Models

An autoregressive model represents the simplest case of typical linear difference equations with constant coefficients. Consider the first-order stochastic linear difference equation or autoregressive model for a zero mean process, $x(t)$, denoted by AR(1), to be defined by

$$x(t) = \alpha_1 x(t - 1) + e(t) \qquad (6.2)$$

where $e(t)$ is an independent, stationary process. The general or pth order AR(p) model can be written as follows

$$x(t) = \alpha_1 x(t - 1) + \alpha_2 x(t - 2) + \ldots + \alpha_p x(t - p) + e(t) \qquad (6.3)$$

where $e(t)$ is an independent, stationary process and p is the number of lags of the dependent variable. Identification, estimation, and forecasting procedures for this model can be found in a variety of sources, such as Granger and Newbold (1986) and Terraza (1981).

Moving-Average Models

The moving-average model embodies the dependency found in the residuals by forming them into linear combinations. One can think of the moving-average model as the inverse function of f, that is, f^{-1}, associated with (6.1). If $e(t)$ is a process with mean zero and constant variance σ^2, then the moving-average model for $x(t)$ denoted as MA(1) is defined by:

$$x(t) = e(t) + \beta_1 e(t - 1) \qquad (6.4)$$

The corresponding ACF for this model is of the form $\rho(1) = -\beta_1/(1 + \beta_1^2)$ for $k = 1$, $\rho(k) = 0$ for $k > 1$. This implies that although observations one period apart are correlated, observations more than one period apart are uncorrelated. Although all MA models are stationary, MA(1) models must satisfy the condition $|\beta_1| < 1$ for the above process to have an infinite autoregressive representation, AR(∞). This has been termed the invertibility condition for MA models; this concept is important, because an MA(1) model is parsimonious compared to an infinite autoregressive representation and this plays an important role in estimation as well as modeling strategy. The general MA(q) model is defined by

$$x(t) = e(t) + \beta_1 e(t - 1) + \ldots + \beta_q e(t - q) \qquad (6.5)$$

Autoregressive Moving-Average Models

The autoregressive moving-average model (ARMA) incorporates both of the above modeling features into (6.6). Occasionally this combination leads to a much simpler time series model that has greater explanatory power than what could be achieved with either higher order AR or MA models. The general specification of an ARMA(p,q) features lag order p for the autoregressive component and lag order q for the moving-average component. Define the general ARMA(1,1) model by

$$x(t) = \alpha_1 x(t - 1) + e(t) + \beta_1 e(t - 1) \qquad (6.6)$$

Then the general ARMA(p,q) specification can be specified as

$$x(t) = \alpha_1 x(t - 1) + \ldots + \alpha_p x(t - p) + e(t) \qquad (6.7)$$

$$+ \beta_1 e(t - 1) + \ldots + \beta_q e(t - q)$$

Order of Integration

Many of the time series data dealt with in political science, economics, and other social sciences are *nonstationary*. For example, notice in the graph in Figure 1.2, that the mean of the party identification time series seems to be changing. That is, if we examine the observations between 1965 and 1975, and the observations between 1980 and 1987, it is quite clear that the value of the mean has changed, that is, the process is nonstationary. Remember that the most fundamental assumption invoked in time series analysis is that the time series be stationary. Therefore, the variable in question cannot be used in the models considered in this chapter, unless it can be transformed to be stationary. As explained earlier in Chapter 1, one procedure that is often performed to achieve this result is based on the technique of successive differencing using the lag operator, L. For example, if $[1 - L]^d x(t)$ requires d differences to achieve stationarity, an ARMA model can be fitted to the differenced data and the process $x(t)$ is referred to as *integrated* of order d. The concept of integration is essential to the class of ARMA models better known as ARIMA(p,d,q) models. Here the order of integration enters directly in the model specification process. As an example, if we want to model growth rates of the percentage of voters who are Democrats,

TABLE 6.1
Identification of Linear Model (AR)P

```
IDENT ΔD(t)
Date: 1-23-1993/Time: 9:25
SMPL range: 1953.3 - 1988.4
Number of observations: 142
```

Autocorrelations	Partial Autocorrelations		ac	pac
| ** | .	| ** | .	| 1	−0.123	−0.123
| ** | .	| ** | .	| 2	−0.118	−0.136
| . |*.	| . | .	| 3	0.053	0.020
| . |*.	| . |*.	| 4	0.051	0.047
| . | .	| . | .	| 5	−0.031	−0.009
| . | .	| . | .	| 6	−0.012	−0.008
| .* | .	| .* | .	| 7	−0.098	−0.113
| *** | .	| *** | .	| 8	−0.199	−0.243
| . |*.	| . | .	| 9	0.105	0.019
| . | .	| .* | .	| 10	−0.014	−0.041
| .* | .	| . | .	| 11	−0.054	−0.019
| . |**	| . |***	| 12	0.184	0.201
| .* | .	| . | .	| 13	−0.052	−0.030
| . | .	| . | .	| 14	0.014	0.031
| . | .	| .* | .	| 15	−0.022	−0.096
| . |***	| . |**	| 16	0.208	0.160

```
Box-Pierce Q-Stat 25.51   Prob 0.0614   SE of Correlations 0.084
Ljung-Box  Q-Stat 27.79   Prob 0.0335
```

the first differences of logs, then we use an ARMA model. Alternatively, if we wanted to model the percentage of voters who are Democrats measured in logs, then an ARIMA(p,1,q) model can be used, where 1 represents the magnitude or degree of the required differencing. The degree of differencing needed to produce a stationary series gives some idea of the order of integration of the series. However, one must proceed with caution in using this difference operator and be careful not to over difference.

Example

Step 1. The attempt is to construct an AR(p) model for the variable $\Delta D(t)$ over the period, 1953.3 to 1988.4. As a first step the sample

TABLE 6.2
Estimation of Linear Model (AR)P

```
LS // Dependent Variable is ΔD(t)
Date: 1-23-1993/Time: 9:28
SMPL range: 1957.2 - 1988.4
Number of observations: 127
=================================================================================================
VARIABLE   COEFFICIENT   STD. ERROR    T-STAT.      2-TAIL SIG.
=================================================================================================
ΔD(-1)     -0.1476697    0.0849304    -1.7387137    0.0846
ΔD(-8)     -0.2064025    0.0879009    -2.3481266    0.0205
ΔD(-12)     0.1866295    0.0849286     2.1974884    0.0299
ΔD(-16)     0.1759112    0.0867687     2.0273589    0.0448
ΔD(-2)     -0.1280719    0.0856933    -1.4945382    0.1376
=================================================================================================
R-squared            0.148480   Mean of dependent var  0.000251
Adjusted R-squared   0.120562   SD of dependent var    0.028996
SE of regression     0.027192   Sum of squared resid   0.090206
Log likelihood       280.1603   F-statistic            5.318321
Durbin-Watson stat   1.967456   Prob(F-statistic)      0.000555
=================================================================================================
```

autocorrelations were estimated for $k = 16$ lags (see Table 6.1). The selection of the value of 16 was based on the assumption that the political process consists of intervals of 2, 3, and 4 years, which translates into 8, 12, and 16 quarters. The diagram is from MicroTSP 7.0 and shows significant lags at 1, 2, 8, 12, and 16 quarters. As an alternative to just picking the lags arbitrarily, one can use the order determination measures in Chapter 8.

Step 2. For illustrative purposes, an autoregressive model was fitted to $\Delta D(t)$ with lags at 1, 2, 8, 12, and 16 quarters. The diagram in Table 6.2 illustrates the MicroTSP 7.0 output for this regression.

7. NONLINEAR MODEL SPECIFICATION

Researchers may fail to construct univariate time series models successfully, if they restrict their specification to linear functions of past observations. In fact, the same past may well contain useful information for the present and future, if nonlinear functions can be discovered. For the dependent stationary time series discussed earlier, a linear representation can occur when the stationary series is also normal. Recently, increased attention has been directed to representations where normality is not present. This has resulted in the specification of a number of different forms of univariate nonlinear time series models (UNLT). Among those considered here are autoregressive conditional heteroscedastic models, bilinear models, threshold autoregressive models, and exponentially autoregressive models.

The generalized Volterra expansion is the logical extension of the ULT (univariate linear time series) models in the previous section. By adding powers of $x(t)$ as well as interaction terms, one would suspect that the function proposed in Equation (6.1) would be better approximated by the inclusion of more terms. The same logic applies here as does a n-order Taylor series expansion used to approximate any polynomial function. Because the ULT models are a first-order Taylor series approximation or a first-order Volterra expansion, a nonlinear version of (6.1) requires higher orders of the Volterra expansion. The ARCH and GARCH models presented below deal with the variance, while the remaining models are based on specification of the mean. Because of space limitations, an example is presented only in the case of the ARCH model.

Autoregressive Conditional Heteroscedastic Models

One of the more popular of the nonlinear time series models is the autoregressive conditional heteroscedastic or ARCH model, which is based on the structure of conditional variances. Recall from Chapter 1 that the concept of weak stationarity required that the mean and the variance of the time series $x(t)$ had to be constant. If, for example, the variance was not constant, it was suggested that transformations from the Box-Cox family could be used to render the variance constant. As an alternative to this transformation process, Engle (1982) has proposed modeling the time-varying variance (i.e., heteroscedasticity) of $x(t)$ directly using the following specification,

$$x(t) = e(t)[h(t)]^{\frac{1}{2}} \qquad (7.1)$$

and

$$h(t) = \alpha_0 + \alpha_1 x^2(t - 1) \qquad (7.2)$$

where $h(t)$ is the conditional variance of $x(t)$. Remember the notion of a general function that reduces the left-hand side of (6.1) to an independent, stationary error process. If we could model the nonstationarity (i.e., nonconstant variance) of the process, $e(t)$, and somehow inject this specification back into the original equation, then the process $e(t)$ could possibly be independent and stationary.

Although the nonlinearity of this model is embodied in the functional term $e(t)x(t-1)$, the addition of the assumption of normality enables the specification to be stated in terms of a linear conditional variance. This conditional variance is based on the information set $I(t-j)$, where $I(t-j)$ represents the information available at time t at lag j. The information set includes causal variables for $x(t)$, that is, $x(t-1)$, $x(t-2)$, ..., and so forth. A more general form of expressions (7.1) and (7.2) using this definition is

$$x(t)|I(t-j) \tag{7.3}$$

which has a normal conditional distribution $N[0,h(t)]$ with zero mean and variance function, $h(t)$, given by

$$h(t) = \alpha_0 + \sum_i \alpha_i e^2(t-i) \quad \text{for } i = 1, \ldots, q \tag{7.4}$$

Expressions (7.3) and (7.4) are known as the autoregressive conditional heteroscedasticity [ARCH(q)] model.

Some of the more important features of this model can be examined through an example. Suppose that we have an AR(1) model with ARCH(1) errors, then

$$x(t) = \alpha_1 x(t-1) + e(t)$$

with conditional variance given by

$$V[e(t)|I(t-1)] = h(t) = \alpha_0 + \alpha_1 e^2(t-1)$$

Assume that $x(t)$ is stationary, which requires that $|\alpha_1| < 1$; in order for the conditional variance to be $h(t) > 0$, then $\alpha_0 > 0$ and $\alpha_1 \geq 0$. For this model the unconditional variance of $e(t)$ is finite in the case that $\alpha_1 < 1$. Although the residuals from this model are serially uncorrelated, they are not independent. This is because the residuals are dependent in their second moments. In order to write an ARCH(q) specification for the residuals, a finite and positive variance must exist if $\sum \alpha_i < 1$ for $i = 1, \ldots, q$.

The conditional variance function does not have to be a linear function of $x^2(t - 1)$, but can have a more generally specific function g such as

$$h(t) = g[x(t - 1), x(t - 2), \ldots, x(t - q), A] \qquad (7.5)$$

where q is the order (number of lags) of the ARCH(q) process and A is a vector of unknown parameters. This specification would be desirable in the case that the conditional moments are nonnormal.

The standard AR-ARCH model can be extended to include other specifications as well. Weiss (1984) assumes that the zero mean process, $x(t)$, is generated not just by an AR(p) process but also by an ARMA (p,q) specification. In this case the conditional variance $h(t)$ has the following specification,

$$h(t) = \alpha_0 + \sum_i \alpha_i e^2(t - i) + \sum_j \varphi_j x^2(t - j) + \varphi_0[x(t) - e(t)]^2 \quad (7.6)$$

where $\varphi_j \geq 0$, $i = 1, \ldots, q$ and $j = 1, \ldots, p$. This is commonly referred to as the ARMA-ARCH model for $x(t)$. The theoretical formulation of (7.6) is more complex then its AR counterpart, but the same logic applies.

Another specification due to Bollerslev (1986, 1988) provides an extension to the above ARCH models by specifying the conditional variance as

$$h(t) = \alpha_0 + \sum_i \alpha_i e^2(t - i) + \sum_j \beta_j h(t - j) \qquad (7.7)$$

where $\beta_j \geq 0$, $i = 1, 2, \ldots, q$ and $j = 1, 2, \ldots, p$. This model integrates the variance into the specification, dictating that the variance be an autoregressive process. This model is referred to as a Generalized ARCH(q,p) or GARCH(q,p) model. Thus, if one believes that changes in the variance are endogenous (that previous variances play a role in determining the current variances), then a GARCH model may be preferred to the ARCH model. It also may be used when the lag structure in an ARCH(q) model is considered to be quite long. A GARCH(q,p) model can be a useful truncation to ARCH processes with a large value for q.

Because it possible that both the mean and variance may be non-constant, Engle, Lilien, and Robbins (1987) extended the ARCH model to allow the conditional variance to affect the mean as well, that is, an ARCH-M (ARCH in mean) model. One also has to consider the case of a unit root in the conditional variance. This is due to the condition mentioned that for a finite and positive variance to exist in an ARCH(q) process, then $\sum_i \alpha_i < 1$ for $i = 1, \ldots, q$. If the summation equals one, then a unit root exists. Please also refer to Engle and Bollerslev (1986) for integrated GARCH (IGARCH) processes of this type.

Bilinear Models

Bilinear time series models are similar to the ARCH models with respect to the specification of the conditional variance. These models were developed by Granger and Andersen (1978) and Subba Rao (1981b) to account for both time varying conditional means and variances. The distinction between the ARCH models and bilinear models lies in the specification of the conditional mean. These models have the capability of explaining "reasonable" nonlinear relationships with a "reasonable" degree of accuracy, although they involve only a limited number of parameters (Brockett, 1976). Such models typically take the following form

$$x(t) + \sum_{j=1}^{p} \alpha_j x(t-j) = \sum_{j=0}^{q} \tau_j e(t-j) + \sum_{i=1}^{m} \sum_{j=1}^{k} \beta_{ij} x(t-i)e(t-j) \qquad (7.8)$$

where the residual process, $e(t)$, is an independent, stationary process. The bilinear form of (7.8) is denoted by BL(p, q, m, k) where p is the autoregressive order; q is the moving-average order; and m and k are the order of the interaction term $x(t - i)e(t - j)$. One can see that an extension of the linear ARMA model to a nonlinear model is performed by including the interaction term $x(t - i)e(t - j)$. This term is included in the model to define the source of the nonlinearity in the time series under investigation. Thus, if $\beta_{ij} = 0$ for all i, j, then (7.8) reduces to a linear ARMA specification.

Because the class of these models is quite large, an important subclass of these models has been considered by Kumar (1986), and Gabr and Subba Rao (1981); they are termed completely bilinear models

$$x(t) = \sum_k \sum_l \beta_{kl} e(t - k) x(t - 1) + e(t) \qquad (7.9)$$

A model of this type is called diagonal if $k = l$; superdiagonal if $k > l$; and subdiagonal if $k < l$. Please refer to Subba Rao (1981b) and Gabr and Subba Rao (1981), who have evaluated the properties of (7.9) including stationarity, invertibility, and the methods of estimation.

Threshold Autoregressive (TAR) Models

Tong and Lim (1980) and Tong (1983) have developed the threshold autoregressive model, a nonlinear extension of the linear autoregressive model, with the added requirement that the coefficients need not be constant. That is, one can begin with a linear model for $x(t)$ and then vary the parameters according to a finite number of past values of $x(t)$ or of some other process. In this case, the coefficients can have more than one mean value. This approach is suitable for specifying cyclical time series models, because they can generate "limit cycle" behavior. For example, the time series can oscillate around a certain value instead of just converging to or diverging from a value. Consider a first-order threshold autoregressive model of the following form, denoted by TAR(2;1,1),

$$x(t) = \begin{cases} \alpha_1 x(t - 1) + e_1(t) & \text{if } x(t - 1) < d \\ \alpha_2 x(t - 1) + e_2(t) & \text{if } x(t - 1) \geq d \end{cases} \qquad (7.10)$$

where $e_1(t)$ and $e_2(t)$ are both independent, stationary processes, α_1 and α_2 are both coefficients, and the constant d is called the threshold or delay parameter.

This specification demonstrates that at some threshold value d, the relationship or dependency between the value of $x(t)$ and its previous lagged values has changed. Given the conditional "if-then statement" shown in Equation (7.10), the specification of the TAR(2;1,1) resembles the specification of two first-order autoregression models that are estimated over two sets. The first set consists of all those observations such that $x(t - 1) < d$, and the other set consists of observations where $x(t - 1) \geq d$. Obviously, Equation (7.10) can be easily extended to the case with S thresholds; that is, we can subdivide or separate the sample and estimate a piecewise autoregressive model based on this chosen subdivision of the sample. Of course, it is also possible to have more lags in (7.10), thereby moving from the TAR(2;1,1) specification to the general TAR(2;p_1,p_2) specification. In general, the TAR(2;p_1,p_2) model can be regarded as a piecewise linear approximation to the general nonlinear pth order autoregressive model.

We should remember that upon selection of S different thresholds, the linear AR model associated with each threshold does not require that the order of these AR(p) models be the same. Different thresholds can have different orders of autoregression. In fact, there is no reason why we cannot extend this threshold model to the more general class of ARMA models, or for that matter, ARIMA models for each threshold specification.

Exponential Autoregressive Models

Ozaki (1980) and Haggan and Ozaki (1981) have introduced the concept of an exponential autoregressive model in an attempt to reproduce certain features of random vibration theory. Their general specification of an exponential autoregressive model [EXPAR(p)] is given by

$$x(t) = \left\{ \varphi_1 + \pi_1 \exp\{-\tau[x(t - 1)]^2\} \right\} x(t - 1) + \ldots \qquad (7.11)$$

$$+ \left\{ \varphi_p + \pi_p \exp\{-\tau[x(t - 1)]^2\} \right\} x(t - p) + e(t)$$

where $e(t)$ is a stationary and independent process and φ_p and π_p are parameters. Notice that if we let α_i represent the term in brackets, then we have a linear autoregressive model of order p where the α_i is an exponential function $x^2(t - 1)$. Therefore, the exponential autoregressive model behaves similarly to that of a threshold autoregressive model, but allows the parameters to change *smoothly* over time. In other words, instead of the abrupt change or "jump" in coefficients expressed by the TAR model, the EXAR model uses the exponential function to facilitate the smoothing of changes over time. Notice that another form of monotonic function could also be selected to achieve smoothing.

TABLE 7.1

Example of an ARCH(1) Model

```
R-SQUARE = 0.1154          R-SQUARE ADJUSTED = 0.0789
VARIANCE OF THE ESTIMATE-SIGMA**2 = 0.73787E-03
STANDARD ERROR OF THE ESTIMATE-SIGMA = 0.27164E-01
SUM OF SQUARED ERRORS-SSE = 0.93710E-01
MEAN OF THE DEPENDENT VARIABLE = 0.25073E-03
LOG OF THE LIKELIHOOD FUNCTION = 277.758
RAW MOMENT R-SQUARE = 0.1155
```

VARIABLE NAME	ESTIMATE COEFFICIENT	ASYMPTOTIC STANDARD ERROR	T-RATIO
$\Delta D(t - 8)$	−0.18510	0.87288E-01	−2.1206
$\Delta D(t - 12)$	0.19888	0.84765E-01	2.3462
$\Delta D(t - 16)$	0.17913	0.86613E-01	2.0682
$\alpha 0$	0.73073E-03	0.10810E-03	6.7596
$\alpha 1$	0.96183E-02	0.78783E-01	0.1221
$\sigma 2$	0.59536E-03	90.726	0.6562E-05

Example

Step 1. The goal is to estimate an ARCH(1) model for the variable $\Delta D(t)$ over the period 1962.1 to 1988.4. An AR model with lags at 8, 12, and 16 with ARCH(1) errors was estimated with the SHAZAM econometric package; the results shown in Table 7.1 were obtained.

8. TESTING FOR MODEL ORDER

Although no single major criterion exists for detecting the order of a time series model, a number of procedures have been put forth. The tests selected are by no means the only ones, and more details on these and other measures and procedures (e.g., R and S arrays, inverse autocorrelations, and corner methods) can be found in de Gooijer, Abraham, Gould, and Robinson (1985).

Testing for model order consists essentially of testing the coefficients of a single regression. For example, if one wants to include another lag into the specification of an AR($p - 1$) model, one can examine the significance of the parameter associated with the pth lag as well as the change in the resulting explained residual variation. This procedure of testing whether to include an additional lag in an econometric relation, is called the Likelihood Ratio Test.

Likelihood Ratio Test (LR)

Assume that the true order of the model is known to be less than k. Anderson (1963, 1971) pointed out that if we assume normality in the residuals of the AR process, then the sequential testing process could be ordered as follows:

(i) Set the null hypothesis, H_0, that the model is AR(k), $k < L$
(ii) Set the alternative hypothesis, H_A, the true order is AR(L)

Construct the Likelihood Ratio test statistic based on the null hypothesis (i) and the alternative hypothesis (ii) to be

$$LR = T\log(\sigma_k^2/\sigma_L^2) \tag{8.1}$$

where σ_k^2 and σ_L^2 are the maximum likelihood estimates of the residuals variance σ_e^2 under the null and alternative hypotheses, respectively. Under the null hypothesis, the test statistic follows a chi-square with $L - k$ degrees of freedom. Remember that the critical values only apply in the case where $x(t)$ or $e(t)$ are normally distributed.

Test Procedure

Step 1. The hypothesis test requires computing the variances of the residual errors in the following two regressions, each having different lag order.

(i) $x(t) = \alpha_0 + \sum_j \alpha_j x(t-j) + e(t)$ for $j = 1, \ldots, p-1$

(ii) $x(t) = \alpha_0 + \sum_j \alpha_j x(t-j) + v(t)$ for $j = 1, \ldots, p$

Step 2. The null and the alternative hypothesis reflect different lag order according to regressions (i) and (ii).

H_0: $x(t)$ is an AR($p - 1$) process
H_A: $x(t)$ is an AR(p) process

The hypothesis test depends on the computation of the likelihood ratio statistic

$$LR = T\log(\sigma_e^2/\sigma_v^2)$$

Step 3. For a given significance level, α, the critical value τ is compared to LR from the table of the chi-square distribution with 1 degree of freedom, $\chi^2(1)$.

Step 4. Reject the null hypothesis in favor of the alternative of one more lag p, if LR > τ.

Example

Step 1. Test the model for the growth rate in party identification variable $\Delta D(t)$ for the period 1953.3 to 1988.4 and $p = 2$, using the following two models, AR(1) and AR(2).

(i) $\Delta D(t) = \alpha_0 + \alpha_1 \Delta D(t - 1) + e(t)$

(ii) $\Delta D(t) = \alpha_0 + \alpha_1 \Delta D(t - 1) + \alpha_2 \Delta D(t - 2) + v(t)$

The residual variance of (i) is $\sigma_e^2 = 0.00078$, and the residual variance of (ii) is $\sigma_v^2 = 0.000767$.

Step 2. The null hypothesis is that $\Delta D(t)$ is AR(1), the test statistic LR becomes

$$LR = (142)\log(0.00078/0.000767) = 1.03649$$

Step 3. For the given significance level, $\alpha = 0.05$, the critical value τ from the chi-square distribution with 2 - 1 degrees of freedom is

$$\tau = 3.84146$$

Step 4. Because LR < τ, one should accept the null hypothesis of an AR(1) model.

One of the main advantages of the LR test is its reliance on classical hypothesis testing procedures that make it uniformly most powerful in a class of the tests that fix the probabilities of accepting a less restrictive hypothesis. However, because the test is sequential, the chosen significance level may not be indicative of the true trade-off between Type I and Type II errors. This is because this procedure is used in the context of only pure AR(p) or MA(q) models, where one has to guard against an unusually large number of lags to approximate the process.

Final Prediction Error (FPE) Test

While the advantage of the likelihood ratio test is that it is the only lag order test explicitly based on statistical distribution theory, de Gooijer et al. (1985) and others have questioned its practicality in the lag order application: "The difficulties in determining the order of an ARMA model by using the Neyman-Pearson approach have introduced the important notion that one should not expect a finite number of observations on a time series process to give a clear cut answer about the true order of the process" (de Gooijer et al., 1985, p. 312).

As a consequence, tests for model order have adapted the concept of forecast error that evaluates the expected one-step-ahead prediction error of a pth order autoregressive process. Akaike (1969, 1970a) was the first to propose such a test, which balanced the errors associated with underestimation of the true order with the errors attributable to overestimation of the true order. This measure is established through the use of a minimization criterion to select order p

$$p = \min\{\text{FPE}(j)|j = 0, \ldots, L\} \tag{8.2}$$

by minimizing over the following FPE ratio

$$\text{FPE}(p) = \sigma_e^2(T + p)/(T - p) \tag{8.3}$$

where $\sigma_e^2 = E\{[x(t + 1) - x^*(t + 1)]^2\}$ is the variance of the difference between the value of $x(t + 1)$ and its forecast value $x^*(t + 1)$ one period ahead. It has been pointed out by de Gooijer et al. (1985) that when $T > p$, the ratio, $(T + p)/(T - p)$, is slightly affected by increases in p, which then suggests that choosing the consistent selection of the correct model order may not be possible.

To circumvent this problem, Akaike (1970b) suggested the following modification to (8.3)

$$\text{FPE}_2(p) = \sigma_e^2 [1 + (p/T^\beta)]/[1 - (p/T)] \quad \text{for } 0 < \beta < 1 \quad (8.4)$$

In addition, McClave (1975) and Bhansali and Downham (1977) suggested

$$\text{FPE}_3(p) = \sigma_e^2 [1 + (\varphi p/T)] \quad \text{for } \varphi > 0 \quad (8.5)$$

The term φ in (8.5) is used to reduce the probability of overestimation of the true order of the model. Howerey (1978) extends the FPE criterion to the case of fitting ARMA models.

Test Procedure

Step 1. Perform the regression on the time series model AR(p) and compute the variance, σ_e^2. The regression is repeated as each value of p is tested.

$$x(t) = \alpha_0 + \sum_j \alpha_j x(t - j) + e(t) \quad \text{for } j = 1, \ldots, p$$

Step 2. Calculate the test ratio. This requires first selecting one form of the three possible variants of the ratio.

(i) $\text{FPE}_1(p) = \sigma_e^2 (T + p)/(T - p)$

(ii) $\text{FPE}_2(p) = \sigma_e^2 [1 + (p/T^\beta)]/[1 - (p/T)] \quad \text{for } 0 < \beta < 1$

(iii) $\text{FPE}_3(p) = \sigma_e^2 [1 + (\varphi p/T)] \quad \text{for } \varphi > 0$

Step 3. The determination of model order p occurs where the value of the FPE ratio for a particular p is at a minimum or does not become lower as other orders of p are tested.

In the case when T is very large compared to p, the ratio in (i) is only slightly affected by changes in p, which makes it difficult to choose the correct model with a high degree of consistency. The modifications introduced by (ii) and (iii) control for this disadvantage. It has also been suggested by Akaike (1979) that FPE tends to overfit. That is, the selected order is larger than the true order. Of course, one has also to identify the upper limit L of the procedure.

Autoregressive Transfer Function (CAT) Test

Parzen (1974) established a criterion that is quite similar to the FPE. The basic distinction between the two is that the autoregressive transfer function (CAT) is based on the technique of spectral analysis that focuses on the frequency of a time series. Using this technique, one can measure the distance between orders in a different manner than in the FPE case. Once again the criterion is to minimize over $p = 0, 1, \ldots, L$, where the test "statistic" is given by

$$CAT(p) = \begin{cases} T^{-1} \sum_j \sigma_e^{-2}(j) - \sigma_e^{-2}(p) & \text{for } p = 1, \ldots, L \text{ and } j = 1, \ldots, p \\ -(1 + T^{-1}) & \text{for } p = 0 \end{cases} \quad (8.6)$$

Tong (1979) offered a modification to (8.6) because it is believed that (8.6) might lead to underestimation of the process; therefore, the range of p is 0 to L and not just from 1 to L

$$CAT(p) = T^{-1} \sum_j \sigma_e^{-2}(j) - \sigma_e^{-2}(p) \quad \text{for } p = 0, 1, \ldots, L \quad (8.7)$$

Test Procedure

Step 1. Perform a regression for $x(t)$ similar to that of FPE and compute the variance. The regression for each value of p tested is

$$x(t) = \alpha_0 + \sum_j \alpha_j x(t - j) + e(t)$$

Step 2. Calculate the $CAT(p)$ ratio for each value of p tested.

$$CAT(p) = \begin{cases} T^{-1}\sum_j \sigma_e^{-2}(j) - \sigma_e^{-2}(p) & \text{for } p = 1, \ldots, L \text{ and } j = 1, \ldots, p \\ -(1 + T^{-1}) & \text{for } p = 0 \end{cases}$$

Step 3. Choose the value of p that minimizes the CAT(p) ratio.

Akaike Information Criterion (AIC) Test

Akaike's (1973) information criterion (AIC) test is a variant of the likelihood ratio method previously discussed. Extending the likelihood ratio concept to the estimation of the ARMA(p,q) models, the AIC test is based on the minimization of

$$\text{AIC}(p,q) = T\log\sigma_e^2 + 2(p + q) \tag{8.8}$$

The first term in (8.8) gives the result due to the fit of the model while the second term delivers a penalty for the inclusion of too high a model order. De Gooijer et al. (1985) state that the AIC can be regarded as a mathematical formulation of the principle of parsimony. One must remember that correct interpretation of (8.8) is only possible when the data are distributed normally.

Bayesian Information Criterion (BIC) Test

The Bayesian information criterion (BIC) test introduces a group of tests that are based on the Bayesian paradigm of statistical analysis. Because of the close similarity of the variants of this test, the test procedure is summarized at the end. Akaike (1979) proposed the BIC based on the selection of the expected posterior loss function for the ARMA(p,q) class of models. The criterion is based on minimizing

$$\text{BIC}(p,q) = T\log\sigma_e^{*2} - (T - p - q)\log[1 - (p + q)/T] \tag{8.9}$$

$$+ (p + q)\log T + (p + q)\log[(p + q)^{-1}(\sigma_x^2/\sigma_e^{*2} - 1)]$$

where σ_x^2 is the variance associated with $x(t)$, and σ_e^{*2} is an estimate of the residual variance. The values of p and q are selected by minimizing

(8.9) for all possible values of p and q. In order to accomplish this, one needs to identify two upper limits L_1 and L_2 for p and q, respectively. This is true for all of the following methods that help to identify the orders of p and q. One must be cautious in this selection.

Bayesian Estimation Method (BEC) Test

Geweke and Meese (1981) have proposed the Bayesian measure

$$BEC(p,q) = \sigma_e^{*2} + (p + q)\sigma_L^2 \log[T/(T - L)] \qquad (8.10)$$

where σ_L^2 is the maximum likelihood estimate of the residual variance of the largest model order. Geweke and Meese have demonstrated that BEC leads to a consistent estimate of model order.

Schwarz (SC) Test

Schwarz (1978) has proposed the following test "statistic" for order selection for an ARMA(p,q) model

$$SC(p,q) = T\log\sigma_e^{*2} + (p + q)\log T \qquad (8.11)$$

where, the values of p and q are selected by minimizing (8.11) over all possible values of p and q.

Hannan-Quinn Criterion Test

As a variant to the Schwarz criterion, Hannan and Quinn (1979) and Hannan (1980) propose the following criterion where one identifies the order of an ARMA(p,q) process by minimizing

$$HQ(p,q) = \log\sigma_e^{*2} + (p + q)c\log[(\log T)/T] \qquad (8.12)$$

where c is a constant to be selected. If the chosen c is larger than 2, Hannan and Quinn have proved that minimization of (8.12) leads to a consistent estimate of the true order.

Test Procedure

Step 1. Perform the regression on the time series model ARMA(p,q) for $x(t)$ for different values of p and q and compute the residual variance in each case

$$x(t) = \alpha_0 + \sum_j \alpha_j x(t-j) + e(t) + \sum_i \beta_i e(t-i)$$

for $i = 1, \ldots, p$ and $j = 1, \ldots, q$

Step 2. Select the test or tests to be used. Compute the value of the test ratios for the criterions selected. Note that all tests employ the same definition of residual variance.

(i) $\text{AIC}(p,q) = T\log\sigma_e^2 + 2(p + q)$

(ii) $\text{BIC}(p,q) = T\log\sigma_e^2 - (T - p - q)\log[1 - (p + q)/T]$

$$+ (p + q)\log T + (p + q)\log[(p + q)^{-1}(\sigma_{x(t)}^2/\sigma_e^2 - 1)]$$

(iii) $\text{BEC}(p,q) = \sigma_e^2 + (p + q)\sigma_L^2\log[T/(T - L)]$ for $p, q = 0, 1, \ldots, L$

(iv) $\text{SC}(p,q) = T\log\sigma_e^2 + (p + q)\log T$

(v) $\text{HQ}(p,q) = \log\sigma_e^2 + (p + q)c\log(\log T)/T$ $c > 2$

Step 3. Choose the values of p and q that minimize the test ratios in each case.

Example

Step 1. For the variable $\Delta D(t)$ over the period 1953.3 to 1988.4, the following two regressions were performed with $p = 1$; an AR(1) model for (i), and $p = 1$ and $q = 1$, ARMA(1,1) for (ii).

$$\text{(i) } \Delta D(t) = \alpha_0 + \alpha_1 \Delta D(t - 1) + e(t)$$

(ii) $\Delta D(t) = \alpha_0 + \alpha_1 \Delta D(t-1) + v(t) + \beta_1 v(t-1)$

The following FPE$_i$, CAT measures are calculated in Step 2a. Step 2b has the order determination measures for the ARMA(p,q) processes.

Step 2a. The following FPE$_i$ and CAT are calculated

(i) $FPE_1(1) = 0.00078(142 + 1)/(142 - 1) = 0.0007916$

(ii) $FPE_2(1) = 0.00078[1 + (1/142)]/[1 - (1/142)] = 0.0007916$

(iii) $FPE_3(1) = 0.00078\{1 + [(1)(1)/142)]\} = 0.0007853$

$CAT(p) = (1/142)(1/0.00078) - (1/0.00078) = -1272.155$

Step 2b. Based on the equation in (ii) of Step 1, we have

(i) $AIC(1,1) = -1010.737$

(ii) $BIC(1,1) = -1014.46$

(iii) $BEC(1,1) = 0.000843$

(iv) $SC(1,1) = -1004.826$

(v) $HQ(1,1) = -7.10085$

Step 3. For illustrative purposes the value of p and q was selected to be one.

The relative quality of these different tests can be summarized as follows. The AIC measure suffers from the problem of inconsistency and overestimation. In addition, the AIC should be used when the sample size is small. Although SC is strongly consistent, it may not be a good approximation to the log of the posterior probability, which may lead to over- or underestimation of true order. The BIC penalizes overparameterization more than the AIC under certain conditions, while the BEC gives consistent estimates of the model order. The HQ also gives consistent estimates of the true order but it leads to underestimation of the true order relative to the AIC for a small number of observations. However, the HQ gives better results than the AIC for a large number of observations.

The main advantage of the Bayesian method is that because one must identify the upper limit on p, this limit can more easily be incorporated into this paradigm, than with the maximum likelihood methods. As it stands, if one wants just to estimate pure AR and MA models, likelihood methods would seem to be preferred as long as the researcher keeps the upper limit L small. Otherwise when fitting ARMA models, it may be a good idea to use all the measures mentioned and to determine how robust the selection of p is to these measures.

9. TESTING THE RESIDUAL PROCESS

A final step in model identification is testing the residual process of the time series model under scrutiny. Not only do such tests help with model identification, but they also help with validation. That is, these tests provide further evidence on the completeness of the resulting goodness of fit. We have observed that the modeling of $x(t)$ results in a residual process $e(t)$, whose behavior may take many different forms. The above tests have been proposed on the time series process $x(t)$ or some transformation of it as well as the residual process $e(t)$. Ideally, because we accounted for all the properties of the raw data, $x(t)$, in the step-by-step testing procedure of the preceding chapters, residual diagnostics might not be needed. However, since the modeling specification process in both the linear and nonlinear domain is largely iterated, the conditions of stationarity and independence of the resulting residuals from these models should be tested. Instead of repeating the procedures mentioned in previous chapters, we will only concern ourselves with three additional tests that are explicitly of the Lagrangian multiplier variety. These "specification" tests are included in this chapter because of their direct reliance on residuals.

Specification Test

As an extension to the typical approach of the Portmanteau tests for examining the independent assumption of the residuals from fitted ARMA models, the Lagrangian Multiplier class of tests require an explicit formulation of an alternative hypothesis. For example, if the null hypothesis is that $x(t)$ is an ARMA(1,2) specification against the alternative hypothesis that $x(t)$ is ARMA(1,3), one only needs to estimate an auxiliary test equation. Godfrey (1979) defines the auxiliary variables of this equation as follows: $z(t)$ and $y(t)$ with $z(t - i)$ defined as the partial derivative of the residuals with respect to the parameters associated with the autoregressive component under the null, $\partial e(t)/\partial \varphi_i$; and $y(t - i)$ defined as the partial derivative of the residuals with respect to the parameters associated with the moving-average component under the null, $\partial e(t)/\partial \theta_i$. Once these auxiliary variables are computed from the hypothesized model estimates, the R^2 from the auxiliary test equation is used to test the specification of the ARMA model.

This score test, as it might be better known, can then be used to determine if the process hypothesized under the null is a correct specification.

Test Procedure

Step 1. Estimate the model under the null hypothesis and calculate the residuals $e(t)$. Formulate the alternative of either ARMA($p+r,q$) or ARMA($p,q+s$). Bear in mind that the alternative ARMA($p+r,q+s$) is not available.

H_0: $x(t)$ is an ARMA(p,q) process
H_A: (i) $x(t)$ is an ARMA($p+r,q$) process $r > 0$
    (ii) $x(t)$ is an ARMA($p,q+s$) process $s > 0$

Step 2. Calculate the auxiliary variables, $z(t)$ and $y(t)$. For $y(t) = z(t)$ = 0 for $t < 0$ and

$$\text{(i) } z(t) = -x(t) + \theta_1 z(t - 1) + \ldots + \theta_q z(t - q)$$

$$\text{(ii) } y(t) = e(t) + \theta_1 y(t - 1) + \ldots + \theta_q y(t - q)$$

Step 3. Perform the following regressions and save the R^2 for each alternative

$$\text{(i) } e(t) = \alpha_1 z(t - 1) + \ldots + \alpha_{p+r} z(t - p - r) + \beta_1 y(t - 1)$$
$$+ \ldots + \beta_q (t - q) + u(t)$$

(ii) $e(t) = \alpha_1 z(t - 1) + \ldots + \alpha_p z(t - p) + \beta_1 y(t - 1)$
$+ \ldots + \beta_{q + s}(t - q - r) + u(t)$

In order to test the null hypothesis, the following test statistic is used

$$LM = TR^2$$

where T is the number of observations and R^2 is the coefficient of determination of the regression conducted in step 1.

Step 4. For the given significance level, α, the critical value τ is based on chi-square distribution with r degrees of freedom, $\chi^2(r)$.

Example

Step 1. For the example variable $D(t)$ over the period 1953.3 to 1988.4, set the following set of hypotheses

H_0: $\Delta D(t)$ is ARMA(0,1)
H_A: $\Delta D(t)$ is ARMA(0,2)

Step 2. Calculate the auxiliary variable $y(t)$; $z(t)$ is not needed because $p = 0$. Perform the regression

$$y(0) = 0, \; y(1) = e(1)$$

$$y(t) = e(t) - (-0.12269)y(t - 1)$$

Step 3. Perform the regression

$$e(t) = \beta_1 y(t - 1) + \beta_2 y(t - 2) + u(t)$$

$$LM = (140)(0.0011864) = 0.166096$$

Step 4. For the significance level, $\alpha = 0.05$, obtain the critical value τ from the table for $\chi^2(1)$.

$$\tau = 3.8416$$

Step 5. Because LM < τ accept the null hypothesis that $\Delta D(t)$ is an ARMA(0,1) process.

ARCH Test

The purpose of performing the ARCH test on the residuals is to test for the presence of heteroscedastic errors in the model. As a consequence the standard errors associated with the parameters in the model will be biased downward. In addition, Weiss (1984) concluded that ignoring the ARCH effect will inevitably result in identifying ARMA models containing too many parameters, that is, overparameterization.

Given the formulation of the ARCH model presented in Chapter 7, it is logical to examine the autocorrelation structure of the squared residuals from the proposed model. This lends itself to the Portmanteau test discussed in the chapter on testing for nonlinear dependence, the McLeod-Li test. As an alternative to this procedure, Engle (1982) suggests a Lagrangian multiplier test, which is presented below.

Test Procedure

Step 1. Select the time series model of interest for $x(t)$ and perform the regression and calculate the residuals $e(t)$. Then perform the second regression, of the following form, for a selected value of p.

$$e^2(t) = \alpha_0 + \sum_i \alpha_i e^2(t - i) + v(t) \quad \text{for } i = 1, \ldots, p$$

Step 2. Establish the null hypothesis of no ARCH effect.

H_0: No ARCH effect
H_A: ARCH effect present

In order to test the null hypothesis, the following test statistic is used

$$\text{LM} = TR^2$$

where T is the number of observations and R^2 is the coefficient of determination of the regression conducted in Step 1.

Step 3. For the given significance level, α, the critical value τ is based on chi-square distribution with p degrees of freedom, $\chi^2(p)$.

Example

Step 1. Select the time series model of interest for $\Delta D(t)$, in this case an AR(1), and calculate the residuals $e(t)$. Then perform the second regression for a selected value of p.

$$e^2(t) = 0.00075 + 0.05129e^2(t-1) + v(t)$$

Step 2. Establish the null hypothesis of no ARCH effect.

H_0: No ARCH effect
H_A: ARCH effect present

In order to test the null hypothesis, the following test statistic is used

$$LM = (142)(0.0026) = 0.37390$$

where T is the number of observations and R^2 is the coefficient of determination of the regression conducted in Step 1.

Step 3. For the given significance level, $\alpha = 0.05$, the critical value τ is based on chi-square distribution with p degrees of freedom, $\chi^2(1)$.

$$\tau = 3.84146$$

Step 4. Because LM $< \tau$, accept the null hypothesis of no ARCH effect.

Bilinear Test

Because the ARCH effect discussed above is primarily based on the notion that the nonlinearity in the residuals is of the multiplicative variety, it is also important to examine the presence of an additive nonlinearity effect. The bilinear test accomplishes this task. The same Lagrangian Multiplier test procedure as in the ARCH effect test is used.

Test Procedure

Step 1. Select the time series model of interest and perform the test and calculate the residuals $e(t)$. Then perform the following regression for a selected value of p.

$$e(t) = \alpha_0 + \sum_i \alpha_i x(t-i) + \sum_i \sum_j \varphi_{ij} x(t-i)e(t-j) + w(t)$$

for $i = 1, \ldots, p$ and $j = 1, \ldots, q$

Step 2. Establish the null hypothesis of no ARCH effect.

H_0: No bilinear effect
H_A: Bilinear effect present

In order to test the null hypothesis, the following test statistic is used

$$LM = TR^2$$

where T is the number of observations and R^2 is the coefficient of determination of the regression conducted in Step 1.

Step 3. For the given significance level, α, the critical value τ is based on chi-square distribution with pq degrees of freedom, $\chi^2(pq)$.

Step 4. If $LM > \tau$, then reject the null hypothesis of the presence of a bilinear effect.

Example

Step 1. Select the time series model of interest for the party identification variable $\Delta D(t)$, in this case an AR(1) model, and calculate the residuals $e(t)$. Then perform the second regression for a selected value of p.

$$e(t) = \alpha_0 + \alpha_1 \Delta D(t-1) + \varphi_{11} \Delta D(t-1)e(t-1) + w(t)$$

Step 2. Establish the null hypothesis of no ARCH effect.

H_0: No bilinear effect

H_A: Bilinear effect present

In order to test the null hypothesis, the following test statistic is used

$$LM = (141)(0.003932) = 0.562872$$

where T is the number of observations and R^2 is the coefficient of determination of the second regression conducted in Step 1.

Step 3. For the given significance level, $\alpha = 0.05$, the critical value τ is based on chi-square distribution with 1 degree of freedom, $\chi^2(1)$.

$$\tau = 3.8416$$

Step 4. Because LM $< \tau$, accept the null hypothesis that there is no bilinear effect.

10. COMPUTATIONAL METHODS FOR PERFORMING THE TESTS

The purpose of this chapter is to provide some insights into the software that is available for constructing the models and performing the tests proposed here. To better facilitate this endeavor, we suggest software such as MicroTSP, RATS, and SHAZAM, which can perform some of the identification tests and the estimation of the regression models associated with each chapter. In Table 10.1, a breakdown has been provided as to which software can best perform a particular test of interest.

Graphical analysis and the sample statistics such as the autocorrelation functions can be found in chapters 2, 7, and 15 of MicroTSP 7.0; most other econometric packages also perform this function.

The Dickey-Fuller tests can easily be done in any regression package. Chapter 16 of MicroTSP 7.0 describes a simple computational method. Other packages such as RATS 3.0 can easily be programmed for these stationarity tests. The Jarque-Bera test for normality can be performed in MicroTSP 7.0. Tests of independence can be done in RATS 3.0, MicroTSP 7.0, and software that enables users to write routines based

TABLE 10.1
Microcomputer Software for Performing Time Series Tests

Test	MicroTSP[a]	RATS[b]	SHAZAM[c]
DF	Y	Y	Y
ADF	Y	Y	Y
Skewness	V	V	V
Kurtosis	V	V	V
Jarque-Bera	Y	V	V
Student Range	V	V	V
Ljung-Box	Y	Y	Y
Box-Pierce	Y	Y	Y
Turning Point	N	V	V
Runs Test	N	V	V
RV Von Neumann	V	V	V
BDS	N	N	N
Keenan	V	V	V
Tsay	V	V	V
LM	V	V	V
McLeod-Li	V	V	V
Hsieh	N	V	V
Likelihood Ratio	V	V	V
FPE	V	V	V
CAT	V	V	V
AIC	V	V	V
BIC	V	V	V
BEC	V	V	V
SC	V	V	V
HQ	V	V	V
Specification	V	V	V
ARCH Test	Y	V	V
Bilinear Test	V	V	V
AR(P)	Y	Y	Y
MA(Q)	Y	Y	Y
ARMA(P,Q)	Y	Y	Y
ARCH(Q)	N	N	Y
GARCH(P,Q)	N	N	Y
BL(P,Q,M,K)	N	N	V
TAR(L,P1,P2)	N	V	N
EXPAR(P)	V	V	V

NOTE: Y stands for a feature of the package. V suggests that the estimation can be performed but the test is not a feature in the software package. The test can be done, however, with auxiliary work outside the package. N means that the test cannot be performed with the package.
a.MicroTSP, Quantitative Micro Software, 4521 Campus Drive, Suite 336, Irvine, CA 92715.
b.RATS, VAR Econometrics, 134 Prospect Avenue So., Minneapolis, MN 55419.
c.SHAZAM, UBC Economics, No. 997-1873 East Mall, Vancouver, BC, V6T-1Z1, Canada.

on estimation results. The BDS test was performed using the software available from one of the developers of the BDS test. Testing for linearity also cannot be easily performed with the software listed above. The McLeod-Li test was performed using MicroTSP 7.0. Most applications of the test statistics detailed in this section come from user-written software. Some of the regression diagnostics mentioned can easily be performed in MicroTSP; other packages require user-written routines. Likelihood ratio and Lagrangian multiplier test of most kinds can be performed in MicroTSP 7.0.

The estimation and identification of linear time series models can easily be performed in the standard statistical packages. For example, the MicroTSP 7.0 command for obtaining the ACF and PACF for a time series is IDENT (variable name). In the case of nonlinear time series models, software is almost nonexistent in the form of user friendly packages. If one is to use a programmable package such as RATS 3.0, then one must write code to perform some of the models. Most often the software for these models is written in FORTRAN and is only available from the model developers. However, through some ingenuity with MicroTSP, one can obtain approximate parameter estimates for the models in Chapter 7.

APPENDIX

TABLE A.1
Critical Values for the Dickey-Fuller Test Statistics

Sample Size				Significance Level				
T	0.01	0.025	0.05	0.10	0.90	0.95	0.975	0.99
No Constant Included in (2.1),τ_1								
25	−2.66	−2.26	−1.95	−1.60	0.92	1.33	1.70	2.16
50	−2.62	−2.25	−1.95	−1.61	0.91	1.31	1.66	2.08
100	−2.60	−2.24	−1.95	−1.61	0.90	1.29	1.64	2.03
250	−2.58	−2.23	−1.95	−1.62	0.89	1.29	1.63	2.01
300	−2.58	−2.23	−1.95	−1.62	0.89	1.28	1.62	2.00
∞	−2.58	−2.23	−1.95	−1.62	0.89	1.28	1.62	2.00
Constant Included in (2.1),τ_2								
25	−3.75	−3.33	−3.00	−2.62	−0.37	0.00	0.34	0.72
50	−3.58	−3.22	−2.93	−2.60	−0.40	−0.03	0.29	0.66
100	−3.51	−3.17	−2.89	−2.58	−0.42	−0.05	0.26	0.63
250	−3.46	−3.14	−2.88	−2.57	−0.42	−0.06	0.24	0.62
300	−3.44	−3.13	−2.87	−2.57	−0.43	−0.07	0.24	0.61
∞	−3.43	−3.12	−2.86	−2.57	−0.44	−0.07	0.23	0.60
Constant and Linear Trend Included in (2.1),τ_3								
25	−4.38	−3.95	−3.60	−3.24	−1.14	−0.80	−0.50	−0.15
50	−4.15	−3.80	−3.50	−3.18	−1.19	−0.87	−0.58	−0.24
100	−4.04	−3.73	−3.45	−3.15	−1.22	−0.90	−0.62	−0.28
250	−3.99	−3.69	−3.43	−3.13	−1.23	−0.92	−0.64	−0.31
300	−3.98	−3.68	−3.42	−3.13	−1.24	−0.93	−0.65	−0.32
∞	−3.96	−3.66	−3.41	−3.12	−1.25	−0.94	−0.66	−0.33

SOURCE: This table was constructed by David A. Dickey using Monte Carlo Methods. Standard errors of the estimates vary, but most are less than 0.02. The table is reproduced from Fuller (1976).

80

TABLE B-1

Areas Under the Standardized Normal Distribution

Example

Pr $(0 \le z \le 1.96) = 0.4750$

Pr $(z \ge 1.96) = 0.5 - 0.4750 = 0.025$

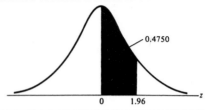

z	.00	.01	.02	.03	.04	.05	.06	.07	.08	.09
0.0	.0000	.0040	.0080	.0120	.0160	.0199	.0239	.0279	.0319	.0359
0.1	.0398	.0438	.0478	.0517	.0557	.0596	.0636	.0675	.0714	.0753
0.2	.0793	.0832	.0871	.0910	.0948	.0987	.1026	.1064	.1103	.1141
0.3	.1179	.1217	.1255	.1293	.1331	.1368	.1406	.1443	.1480	.1517
0.4	.1554	.1591	.1628	.1664	.1700	.1736	.1772	.1808	.1844	.1879
0.5	.1915	.1950	.1985	.2019	.2054	.2088	.2123	.2157	.2190	.2224
0.6	.2257	.2291	.2324	.2357	.2389	.2422	.2454	.2486	.2517	.2549
0.7	.2580	.2611	.2642	.2673	.2704	.2734	.2764	.2794	.2823	.2852
0.8	.2881	.2910	.2939	.2967	.2995	.3023	.3051	.3078	.3106	.3133
0.9	.3159	.3186	.3212	.3238	.3264	.3289	.3315	.3340	.3365	.3389
1.0	.3413	.3438	.3461	.3485	.3508	.3531	.3554	.3577	.3599	.3621
1.1	.3643	.3665	.3686	.3708	.3729	.3749	.3770	.3790	.3810	.3830
1.2	.3849	.3869	.3888	.3907	.3925	.3944	.3962	.3980	.3997	.4015
1.3	.4032	.4049	.4066	.4082	.4099	.4115	.4131	.4147	.4162	.4177
1.4	.4192	.4207	.4222	.4236	.4251	.4265	.4279	.4292	.4306	.4319
1.5	.4332	.4345	.4357	.4370	.4382	.4394	.4406	.4418	.4429	.4441
1.6	.4452	.4463	.4474	.4484	.4495	.4505	.4515	.4525	.4535	.4545
1.7	.4554	.4564	.4573	.4582	.4591	.4599	.4608	.4616	.4625	.4633
1.8	.4641	.4649	.4656	.4664	.4671	.4678	.4686	.4693	.4699	.4706
1.9	.4713	.4719	.4726	.4732	.4738	.4744	.4750	.4756	.4761	.4767
2.0	.4772	.4778	.4783	.4788	.4793	.4798	.4803	.4808	.4812	.4817
2.1	.4821	.4826	.4830	.4834	.4838	.4842	.4846	.4850	.4854	.4857
2.2	.4861	.4864	.4868	.4871	.4875	.4878	.4881	.4884	.4887	.4890
2.3	.4893	.4896	.4898	.4901	.4904	.4906	.4909	.4911	.4913	.4916
2.4	.4918	.4920	.4922	.4925	.4927	.4929	.4931	.4932	.4934	.4936
2.5	.4938	.4940	.4941	.4943	.4945	.4946	.4948	.4949	.4951	.4952
2.6	.4953	.4955	.4956	.4957	.4959	.4960	.4961	.4962	.4963	.4964
2.7	.4965	.4966	.4967	.4968	.4969	.4970	.4971	.4972	.4973	.4974
2.8	.4974	.4975	.4976	.4977	.4977	.4978	.4979	.4979	.4980	.4981
2.9	.4981	.4982	.4982	.4983	.4984	.4984	.4985	.4985	.4986	.4986
3.0	.4987	.4987	.4987	.4988	.4988	.4989	.4989	.4989	.4990	.4990

Note: This table gives the area in the right-hand tail of the distribution (i.e., $z \ge 0$). But since the normal distribution is symmetrical about $z = 0$, the area in the left-hand tail is the same as the area in the corresponding right-hand tail. For example, $P(-1.96 \le z \le 0) = 0.4750$. Therefore, $P(-1.96 \le z \le 1.96) = 2(0.4750) = 0.95$.

Source: Table values for Tables B-1 to B-4 are from E. S. Pearson and H. O. Hartley, (eds.). *Biometrika Tables for Statisticians* (vol 1, 3d. ed.). Cambridge University Press, New York, 1966. Reproduced by permission of the editors and trustees of *Biometrika*. Table example and format is from D. Gujarati, *Essentials of Econometrics*, McGraw-Hill, Inc., 1992. Reproduced by permission of McGraw-Hill, Inc.

TABLE B-2

Percentage Points of the t Distribution

Example

$\Pr(t > 2.086) = 0.025$

$\Pr(t > 1.725) = 0.05 \qquad \text{for df} = 20$

$\Pr(|t| > 1.725) = 0.10$

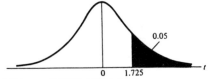

Pr df	0.25 0.50	0.10 0.20	0.05 0.10	0.025 0.05	0.01 0.02	0.005 0.010	0.001 0.002
1	1.000	3.078	6.314	12.706	31.821	63.657	318.31
2	0.816	1.886	2.920	4.303	6.965	9.925	22.327
3	0.765	1.638	2.353	3.182	4.541	5.841	10.214
4	0.741	1.533	2.132	2.776	3.747	4.604	7.173
5	0.727	1.476	2.015	2.571	3.365	4.032	5.893
6	0.718	1.440	1.943	2.447	3.143	3.707	5.208
7	0.711	1.415	1.895	2.365	2.998	3.499	4.785
8	0.706	1.397	1.860	2.306	2.896	3.355	4.501
9	0.703	1.383	1.833	2.262	2.821	3.250	4.297
10	0.700	1.372	1.812	2.228	2.764	3.169	4.144
11	0.697	1.363	1.796	2.201	2.718	3.106	4.025
12	0.695	1.356	1.782	2.179	2.681	3.055	3.930
13	0.694	1.350	1.771	2.160	2.650	3.012	3.852
14	0.692	1.345	1.761	2.145	2.624	2.977	3.787
15	0.691	1.341	1.753	2.131	2.602	2.947	3.733
16	0.690	1.337	1.746	2.120	2.583	2.921	3.686
17	0.689	1.333	1.740	2.110	2.567	2.898	3.646
18	0.688	1.330	1.734	2.101	2.552	2.878	3.610
19	0.688	1.328	1.729	2.093	2.539	2.861	3.579
20	0.687	1.325	1.725	2.086	2.528	2.845	3.552
21	0.686	1.323	1.721	2.080	2.518	2.831	3.527
22	0.686	1.321	1.717	2.074	2.508	2.819	3.505
23	0.685	1.319	1.714	2.069	2.500	2.807	3.485
24	0.685	1.318	1.711	2.064	2.492	2.797	3.467
25	0.684	1.316	1.708	2.060	2.485	2.787	3.450
26	0.684	1.315	1.706	2.056	2.479	2.779	3.435
27	0.684	1.314	1.703	2.052	2.473	2.771	3.421
28	0.683	1.313	1.701	2.048	2.467	2.763	3.408
29	0.683	1.311	1.699	2.045	2.462	2.756	3.396
30	0.683	1.310	1.697	2.042	2.457	2.750	3.385
40	0.681	1.303	1.684	2.021	2.423	2.704	3.307
60	0.679	1.296	1.671	2.000	2.390	2.660	3.232
120	0.677	1.289	1.658	1.980	2.358	2.167	3.160
∞	0.674	1.282	1.645	1.960	2.326	2.576	3.090

Note: The smaller probability shown at the head of each column is the area in one tail; the larger probability is the area in both tails.

TABLE B-3
Upper Percentage Points of the F Distribution

Example

$\Pr(F > 1.59) = 0.25$
$\Pr(F > 2.42) = 0.10$ for df $N_1 = 10$
$\Pr(F > 3.14) = 0.05$ and $N_2 = 9$
$\Pr(F > 5.26) = 0.01$

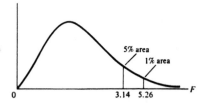

df for denom- inator N_2	Pr	1	2	3	4	5	6	7	8	9	10	11	12
1	.25	5.83	7.50	8.20	8.58	8.82	8.98	9.10	9.19	9.26	9.32	9.36	9.41
	.10	39.9	49.5	53.6	55.8	57.2	58.2	58.9	59.4	59.9	60.2	60.5	60.7
	.05	161	200	216	225	230	234	237	239	241	242	243	244
2	.25	2.57	3.00	3.15	3.23	3.28	3.31	3.34	3.35	3.37	3.38	3.39	3.39
	.10	8.53	9.00	9.16	9.24	9.29	9.33	9.35	9.37	9.38	9.39	9.40	9.41
	.05	18.5	19.0	19.2	19.2	19.3	19.3	19.4	19.4	19.4	19.4	19.4	19.4
	.01	98.5	99.0	99.2	99.2	99.3	99.3	99.4	99.4	99.4	99.4	99.4	99.4
3	.25	2.02	2.28	2.36	2.39	2.41	2.42	2.43	2.44	2.44	2.44	2.45	2.45
	.10	5.54	5.46	5.39	5.34	5.31	5.28	5.27	5.25	5.25	5.24	5.23	5.22
	.05	10.1	9.55	9.28	9.12	9.01	8.94	8.89	8.85	8.81	8.79	8.76	8.74
	.01	34.1	30.8	29.5	28.7	28.2	27.9	27.7	27.5	27.3	27.2	27.1	27.1
4	.25	1.81	2.00	2.05	2.06	2.07	2.08	2.08	2.08	2.08	2.08	2.08	2.08
	.10	4.54	4.32	4.19	4.11	4.05	4.01	3.98	3.95	3.94	3.92	3.91	3.90
	.05	7.71	6.94	6.59	6.39	6.26	6.16	6.09	6.04	6.00	5.96	5.94	5.91
	.01	21.2	18.0	16.7	16.0	15.5	15.2	15.0	14.8	14.7	14.5	14.4	14.4
5	.25	1.69	1.85	1.88	1.89	1.89	1.89	1.89	1.89	1.89	1.89	1.89	1.89
	.10	4.06	3.78	3.62	3.52	3.45	3.40	3.37	3.34	3.32	3.30	3.28	3.27
	.05	6.61	5.79	5.41	5.19	5.05	4.95	4.88	4.82	4.77	4.74	4.71	4.68
	.01	16.3	13.3	12.1	11.4	11.0	10.7	10.5	10.3	10.2	10.1	9.96	9.89
6	.25	1.62	1.76	1.78	1.79	1.79	1.78	1.78	1.78	1.77	1.77	1.77	1.77
	.10	3.78	3.46	3.29	3.18	3.11	3.05	3.01	2.98	2.96	2.94	2.92	2.90
	.05	5.99	5.14	4.76	4.53	4.39	4.28	4.21	4.15	4.10	4.06	4.03	4.00
	.01	13.7	10.9	9.78	9.15	8.75	8.47	8.26	8.10	7.98	7.87	7.79	7.72
7	.25	1.57	1.70	1.72	1.72	1.71	1.71	1.70	1.70	1.69	1.69	1.69	1.68
	.10	3.59	3.26	3.07	2.96	2.88	2.83	2.78	2.75	2.72	2.70	2.68	2.67
	.05	5.59	4.74	4.35	4.12	3.97	3.87	3.79	3.73	3.68	3.64	3.60	3.57
	.01	12.2	9.55	8.45	7.85	7.46	7.19	6.99	6.84	6.72	6.62	6.54	6.47
8	.25	1.54	1.66	1.67	1.66	1.66	1.65	1.64	1.64	1.63	1.63	1.63	1.62
	.10	3.46	3.11	2.92	2.81	2.73	2.67	2.62	2.59	2.56	2.54	2.52	2.50
	.05	5.32	4.46	4.07	3.84	3.69	3.58	3.50	3.44	3.39	3.35	3.31	3.28
	.01	11.3	8.65	7.59	7.01	6.63	6.37	6.18	6.03	5.91	5.81	5.73	5.67
9	.25	1.51	1.62	1.63	1.63	1.62	1.61	1.60	1.60	1.59	1.59	1.58	1.58
	.10	3.36	3.01	2.81	2.69	2.61	2.55	2.51	2.47	2.44	2.42	2.40	2.38
	.05	5.12	4.26	3.86	3.63	3.48	3.37	3.29	3.23	3.18	3.14	3.10	3.07
	.01	10.6	8.02	6.99	6.42	6.06	5.80	5.61	5.47	5.35	5.26	5.18	5.11

			df for numerator N_1										df for denominator N_2
15	**20**	**24**	**30**	**40**	**50**	**60**	**100**	**120**	**200**	**500**	**∞**	**Pr**	**N_2**
9.49	9.58	9.63	9.67	9.71	9.74	9.76	9.78	9.80	9.82	9.84	9.85	.25	
61.2	61.7	62.0	62.3	62.5	62.7	62.8	63.0	63.1	63.2	63.3	63.3	.10	1
246	248	249	250	251	252	252	253	253	254	254	254	.05	
3.41	3.43	3.43	3.44	3.45	3.45	3.46	3.47	3.47	3.48	3.48	3.48	.25	
9.42	9.44	9.45	9.46	9.47	9.47	9.47	9.48	9.48	9.49	9.49	9.49	.10	2
19.4	19.4	19.5	19.5	19.5	19.5	19.5	19.5	19.5	19.5	19.5	19.5	.05	
99.4	99.4	99.5	99.5	99.5	99.5	99.5	99.5	99.5	99.5	99.5	99.5	.01	
2.46	2.46	2.46	2.47	2.47	2.47	2.47	2.47	2.47	2.47	2.47	2.47	.25	
5.20	5.18	5.18	5.17	5.16	5.15	5.15	5.14	5.14	5.14	5.14	5.13	.10	3
8.70	8.66	8.64	8.62	8.59	8.58	8.57	8.55	8.55	8.54	8.53	8.53	.05	
26.9	26.7	26.6	26.5	26.4	26.4	26.3	26.2	26.2	26.2	26.1	26.1	.01	
2.08	2.08	2.08	2.08	2.08	2.08	2.08	2.08	2.08	2.08	2.08	2.08	.25	
3.87	3.84	3.83	3.82	3.80	3.80	3.79	3.78	3.78	3.77	3.76	3.76	.10	4
5.86	5.80	5.77	5.75	5.72	5.70	5.69	5.66	5.66	5.65	5.64	5.63	.05	
14.2	14.0	13.9	13.8	13.7	13.7	13.7	13.6	13.6	13.5	13.5	13.5	.01	
1.89	1.88	1.88	1.88	1.88	1.88	1.87	1.87	1.87	1.87	1.87	1.87	.25	
3.24	3.21	3.19	3.17	3.16	3.15	3.14	3.13	3.12	3.12	3.11	3.10	.10	5
4.62	4.56	4.53	4.50	4.46	4.44	4.43	4.41	4.40	4.39	4.37	4.36	.05	
9.72	9.55	9.47	9.38	9.29	9.24	9.20	9.13	9.11	9.08	9.04	9.02	.01	
1.76	1.76	1.75	1.75	1.75	1.75	1.74	1.74	1.74	1.74	1.74	1.74	.25	
2.87	2.84	2.82	2.80	2.78	2.77	2.76	2.75	2.74	2.73	2.73	2.72	.10	6
3.94	3.87	3.84	3.81	3.77	3.75	3.74	3.71	3.70	3.69	3.68	3.67	.05	
7.56	7.40	7.31	7.23	7.14	7.09	7.06	6.99	6.97	6.93	6.90	6.88	.01	
1.68	1.67	1.67	1.66	1.66	1.66	1.65	1.65	1.65	1.65	1.65	1.65	.25	
2.63	2.59	2.58	2.56	2.54	2.52	2.51	2.50	2.49	2.48	2.48	2.47	.10	7
3.51	3.44	3.41	3.38	3.34	3.32	3.30	3.27	3.27	3.25	3.24	3.23	.05	
6.31	6.16	6.07	5.99	5.91	5.86	5.82	5.75	5.74	5.70	5.67	5.65	.10	
1.62	1.61	1.60	1.60	1.59	1.59	1.59	1.58	1.58	1.58	1.58	1.58	.25	
2.46	2.42	2.40	2.38	2.36	2.35	2.34	2.32	2.32	2.31	2.30	2.29	.10	8
3.22	3.15	3.12	3.08	3.04	2.02	3.01	2.97	2.97	2.95	2.94	2.93	.05	
5.52	5.36	5.28	5.20	5.12	5.07	5.03	4.96	4.95	4.91	4.88	4.86	.01	
1.57	1.56	1.56	1.55	1.55	1.54	1.54	1.53	1.53	1.53	1.53	1.53	.25	
2.34	2.30	2.28	2.25	2.23	2.22	2.21	2.19	2.18	2.17	2.17	2.16	.10	9
3.01	2.94	2.90	2.86	2.83	2.80	2.79	2.76	2.75	2.73	2.72	2.71	.05	
4.96	4.81	4.73	4.65	4.57	4.52	4.48	4.42	4.40	4.36	4.33	4.31	.01	

TABLE B-3
Continued

df for denominator N_2	Pr	\multicolumn{12}{c}{df for numerator N_1}											

df for denom-inator N_2	Pr	1	2	3	4	5	6	7	8	9	10	11	12
10	.25	1.49	1.60	1.60	1.59	1.59	1.58	1.57	1.56	1.56	1.55	1.55	1.54
	.10	3.29	2.92	2.73	2.61	2.52	2.46	2.41	2.38	2.35	2.32	2.30	2.28
	.05	4.96	4.10	3.71	3.48	3.33	3.22	3.14	3.07	3.02	2.98	2.94	2.91
	.01	10.0	7.56	6.55	5.99	5.64	5.39	5.20	5.06	4.94	4.85	4.77	4.71
11	.25	1.47	1.58	1.58	1.57	1.56	1.55	1.54	1.53	1.53	1.52	1.52	1.51
	.10	3.23	2.86	2.66	2.54	2.45	2.39	2.34	2.30	2.27	2.25	2.23	2.21
	.05	4.84	3.98	3.59	3.36	3.20	3.09	3.01	2.95	2.90	2.85	2.82	2.79
	.01	9.65	7.21	6.22	5.67	5.32	5.07	4.89	4.74	4.63	4.54	4.46	4.40
12	.25	1.46	1.56	1.56	1.55	1.54	1.53	1.52	1.51	1.51	1.50	1.50	1.49
	.10	3.18	2.81	2.61	2.48	2.39	2.33	2.28	2.24	2.21	2.19	2.17	2.15
	.05	4.75	3.89	3.49	3.26	3.11	3.00	2.91	2.85	2.80	2.75	2.72	2.69
	.01	9.33	6.93	5.95	5.41	5.06	4.82	4.64	4.50	4.39	4.30	4.22	4.16
13	.25	1.45	1.55	1.55	1.53	1.52	1.51	1.50	1.49	1.49	1.48	1.47	1.47
	.10	3.14	2.76	2.56	2.43	2.35	2.28	2.23	2.20	2.16	2.14	2.12	2.10
	.05	4.67	3.81	3.41	3.18	3.03	2.92	2.83	2.77	2.71	2.67	2.63	2.60
	.01	9.07	6.70	5.74	5.21	4.86	4.62	4.44	4.30	4.19	4.10	4.02	3.96
14	.25	1.44	1.53	1.53	1.52	1.51	1.50	1.49	1.48	1.47	1.46	1.46	1.45
	.10	3.10	2.73	2.52	2.39	2.31	2.24	2.19	2.15	2.12	2.10	2.08	2.05
	.05	4.60	3.74	3.34	3.11	2.96	2.85	2.76	2.70	2.65	2.60	2.57	2.53
	.01	8.86	6.51	5.56	5.04	4.69	4.46	4.28	4.14	4.03	3.94	3.86	3.80
15	.25	1.43	1.52	1.52	1.51	1.49	1.48	1.47	1.46	1.46	1.45	1.44	1.44
	.10	3.07	2.70	2.49	2.36	2.27	2.21	2.16	2.12	2.09	2.06	2.04	2.02
	.05	4.54	3.68	3.29	3.06	2.90	2.79	2.71	2.64	2.59	2.54	2.51	2.48
	.01	8.68	6.36	5.42	4.89	4.56	4.32	4.14	4.00	3.89	3.80	3.73	3.67
16	.25	1.42	1.51	1.51	1.50	1.48	1.47	1.46	1.45	1.44	1.44	1.44	1.43
	.10	3.05	2.67	2.46	2.33	2.24	2.18	2.13	2.09	2.06	2.03	2.01	1.99
	.05	4.49	3.63	3.24	3.01	2.85	2.74	2.66	2.59	2.54	2.49	2.46	2.42
	.01	8.53	6.23	5.29	4.77	4.44	4.20	4.03	3.89	3.78	3.69	3.62	3.55
17	.25	1.42	1.51	1.50	1.49	1.47	1.46	1.45	1.44	1.43	1.43	1.42	1.41
	.10	3.03	2.64	2.44	2.31	2.22	2.15	2.10	2.06	2.03	2.00	1.98	1.96
	.05	4.45	3.59	3.20	2.96	2.81	2.70	2.61	2.55	2.49	2.45	2.41	2.38
	.01	8.40	6.11	5.18	4.67	4.34	4.10	3.93	3.79	3.68	3.59	3.52	3.46
18	.25	1.41	1.50	1.49	1.48	1.46	1.45	1.44	1.43	1.42	1.42	1.41	1.40
	.10	3.01	2.62	2.42	2.29	2.20	2.13	2.08	2.04	2.00	1.98	1.96	1.93
	.05	4.41	3.55	3.16	2.93	2.77	2.66	2.58	2.51	2.46	2.41	2.37	2.34
	.01	8.29	6.01	5.09	4.58	4.25	4.01	3.84	3.71	3.60	3.51	3.43	3.37
19	.25	1.41	1.49	1.49	1.47	1.46	1.44	1.43	1.42	1.41	1.41	1.40	1.40
	.10	2.99	2.61	2.40	2.27	2.18	2.11	2.06	2.02	1.98	1.96	1.94	1.91
	.05	4.38	3.52	3.13	2.90	2.74	2.63	2.54	2.48	2.42	2.38	2.34	2.31
	.01	8.18	5.93	5.01	4.50	4.17	3.94	3.77	3.63	3.52	3.43	3.36	3.30
20	.25	1.40	1.49	1.48	1.46	1.45	1.44	1.43	1.42	1.41	1.40	1.39	1.39
	.10	2.97	2.59	2.38	2.25	2.16	2.09	2.04	2.00	1.96	1.94	1.92	1.89
	.05	4.35	3.49	3.10	2.87	2.71	2.60	2.51	2.45	2.39	2.35	2.31	2.28
	.01	8.10	5.85	4.94	4.43	4.10	3.87	3.70	3.56	3.46	3.37	3.29	3.23

				df for numerator N_1									df for denominator
15	20	24	30	40	50	60	100	120	200	500	∞	Pr	N_2
1.53	1.52	1.52	1.51	1.51	1.50	1.50	1.49	1.49	1.49	1.48	1.48	.25	
2.24	2.20	2.18	2.16	2.13	2.12	2.11	2.09	2.08	2.07	2.06	2.06	.10	10
2.85	2.77	2.74	2.70	2.66	2.64	2.62	2.59	2.58	2.56	2.55	2.54	.05	
4.56	4.41	4.33	4.25	4.17	4.12	4.08	4.01	4.00	3.96	3.93	3.91	.01	
1.50	1.49	1.49	1.48	1.47	1.47	1.47	1.46	1.46	1.46	1.45	1.45	.25	
2.17	2.12	2.10	2.08	2.05	2.04	2.03	2.00	2.00	1.99	1.98	1.97	.10	11
2.72	2.65	2.61	2.57	2.53	2.51	2.49	2.46	2.45	2.43	2.42	2.40	.05	
4.25	4.10	4.02	3.94	3.86	3.81	3.78	3.71	3.69	3.66	3.62	3.60	.01	
1.48	1.47	1.46	1.45	1.45	1.44	1.44	1.43	1.43	1.43	1.42	1.42	.25	
2.10	2.06	2.04	2.01	1.99	1.97	1.96	1.94	1.93	1.92	1.91	1.90	.10	12
2.62	2.54	2.51	2.47	2.43	2.40	2.38	2.35	2.34	2.32	2.31	2.30	.05	
4.01	3.86	3.78	3.70	3.62	3.57	3.54	3.47	3.45	3.41	3.38	3.36	.01	
1.46	1.45	1.44	1.43	1.42	1.42	1.42	1.41	1.41	1.40	1.40	1.40	.25	
2.05	2.01	1.98	1.96	1.93	1.92	1.90	1.88	1.88	1.86	1.85	1.85	.10	13
2.53	2.46	2.42	2.38	2.34	2.31	2.30	2.26	2.25	2.23	2.22	2.21	.05	
3.82	3.66	3.59	3.51	3.43	3.38	3.34	3.27	3.25	3.22	3.19	3.17	.01	
1.44	1.43	1.42	1.41	1.41	1.40	1.40	1.39	1.39	1.39	1.38	1.38	.25	
2.01	1.96	1.94	1.91	1.89	1.87	1.86	1.83	1.83	1.82	1.80	1.80	.10	14
2.46	2.39	2.35	2.31	2.27	2.24	2.22	2.19	2.18	2.16	2.14	2.13	.05	
3.66	3.51	3.43	3.35	3.27	3.22	3.18	3.11	3.09	3.06	3.03	3.00	.01	
1.43	1.41	1.41	1.40	1.39	1.39	1.38	1.38	1.37	1.37	1.36	1.36	.25	
1.97	1.92	1.90	1.87	1.85	1.83	1.82	1.79	1.79	1.77	1.76	1.76	.10	15
2.40	2.33	2.29	2.25	2.20	2.18	2.16	2.12	2.11	2.10	2.08	2.07	.05	
3.52	3.37	3.29	3.21	3.13	3.08	3.05	2.98	2.96	2.92	2.89	2.87	.01	
1.41	1.40	1.39	1.38	1.37	1.37	1.36	1.36	1.35	1.35	1.34	1.34	.25	
1.94	1.89	1.87	1.84	1.81	1.79	1.78	1.76	1.75	1.74	1.73	1.72	.10	16
2.35	2.28	2.24	2.19	2.15	2.12	2.11	2.07	2.06	2.04	2.02	2.01	.05	
3.41	3.26	3.18	3.10	3.02	2.97	2.93	2.86	2.84	2.81	2.78	2.75	.01	
1.40	1.39	1.38	1.37	1.36	1.35	1.35	1.34	1.34	1.34	1.33	1.33	.25	
1.91	1.86	1.84	1.81	1.78	1.76	1.75	1.73	1.72	1.71	1.69	1.69	.10	17
2.31	2.23	2.19	2.15	2.10	2.08	2.06	2.02	2.01	1.99	1.97	1.96	.05	
3.31	3.16	3.08	3.00	2.92	2.87	2.83	2.76	2.75	2.71	2.68	2.65	.01	
1.39	1.38	1.37	1.36	1.35	1.34	1.34	1.33	1.33	1.32	1.32	1.32	.25	
1.89	1.84	1.81	1.78	1.75	1.74	1.72	1.70	1.69	1.68	1.67	1.66	.10	18
2.27	2.19	2.15	2.11	2.06	2.04	2.02	1.98	1.97	1.95	1.93	1.92	.05	
3.23	3.08	3.00	2.92	2.84	2.78	2.75	2.68	2.66	2.62	2.59	2.57	.01	
1.38	1.37	1.36	1.35	1.34	1.33	1.33	1.32	1.32	1.31	1.31	1.30	.25	
1.86	1.81	1.79	1.76	1.73	1.71	1.70	1.67	1.67	1.65	1.64	1.63	.10	19
2.23	2.16	2.11	2.07	2.03	2.00	1.98	1.94	1.93	1.91	1.89	1.88	.05	
3.15	3.00	2.92	2.84	2.76	2.71	2.67	2.60	2.58	2.55	2.51	2.49	.01	
1.37	1.36	1.35	1.34	1.33	1.33	1.32	1.31	1.31	1.30	1.30	1.29	.25	
1.84	1.79	1.77	1.74	1.71	1.69	1.68	1.65	1.64	1.63	1.62	1.61	.10	20
2.20	2.12	2.08	2.04	1.99	1.97	1.95	1.91	1.90	1.88	1.86	1.84	.05	
3.09	2.94	2.86	2.78	2.69	2.64	2.61	2.54	2.52	2.48	2.44	2.42	.01	

TABLE B-3
Continued

df for denominator N_2	Pr	\multicolumn{12}{c}{df for numerator N_1}
		1	2	3	4	5	6	7	8	9	10	11	12
22	.25	1.40	1.48	1.47	1.45	1.44	1.42	1.41	1.40	1.39	1.39	1.38	1.37
	.10	2.95	2.56	2.35	2.22	2.13	2.06	2.01	1.97	1.93	1.90	1.88	1.86
	.05	4.30	3.44	3.05	2.82	2.66	2.55	2.46	2.40	2.34	2.30	2.26	2.23
	.01	7.95	5.72	4.82	4.31	3.99	3.76	3.59	3.45	3.35	3.26	3.18	3.12
24	.25	1.39	1.47	1.46	1.44	1.43	1.41	1.40	1.39	1.38	1.38	1.37	1.36
	.10	2.93	2.54	2.33	2.19	2.10	2.04	1.98	1.94	1.91	1.88	1.85	1.83
	.05	4.26	3.40	3.01	2.78	2.62	2.51	2.42	2.36	2.30	2.25	2.21	2.18
	.01	7.82	5.61	4.72	4.22	3.90	3.67	3.50	3.36	3.26	3.17	3.09	3.03
26	.25	1.38	1.46	1.45	1.44	1.42	1.41	1.39	1.38	1.37	1.37	1.36	1.35
	.10	2.91	2.52	2.31	2.17	2.08	2.01	1.96	1.92	1.88	1.86	1.84	1.81
	.05	4.23	3.37	2.98	2.74	2.59	2.47	2.39	2.32	2.27	2.22	2.18	2.15
	.01	7.72	5.53	4.64	4.14	3.82	3.59	3.42	3.29	3.18	3.09	3.02	2.96
28	.25	1.38	1.46	1.45	1.43	1.41	1.40	1.39	1.38	1.37	1.36	1.35	1.34
	.10	2.89	2.50	2.29	2.16	2.06	2.00	1.94	1.90	1.87	1.84	1.81	1.79
	.05	4.20	3.34	2.95	2.71	2.56	2.45	2.36	2.29	2.24	2.19	2.15	2.12
	.01	7.64	5.45	4.57	4.07	3.75	3.53	3.36	3.23	3.12	3.03	2.96	2.90
30	.25	1.38	1.45	1.44	1.42	1.41	1.39	1.38	1.37	1.36	1.35	1.35	1.34
	.10	2.88	2.49	2.28	2.14	2.05	1.98	1.93	1.88	1.85	1.82	1.79	1.77
	.05	4.17	3.32	2.92	2.69	2.53	2.42	2.33	2.27	2.21	2.16	2.13	2.09
	.01	7.56	5.39	4.51	4.02	3.70	3.47	3.30	3.17	3.07	2.98	2.91	2.84
40	.25	1.36	1.44	1.42	1.40	1.39	1.37	1.36	1.35	1.34	1.33	1.32	1.31
	.10	2.84	2.44	2.23	2.09	2.00	1.93	1.87	1.83	1.79	1.76	1.73	1.71
	.05	4.08	3.23	2.84	2.61	2.45	2.34	2.25	2.18	2.12	2.08	2.04	2.00
	.01	7.31	5.18	4.31	3.83	3.51	3.29	3.12	2.99	2.89	2.80	2.73	2.66
60	.25	1.35	1.42	1.41	1.38	1.37	1.35	1.33	1.32	1.31	1.30	1.29	1.29
	.10	2.79	2.39	2.18	2.04	1.95	1.87	1.82	1.77	1.74	1.71	1.68	1.66
	.05	4.00	3.15	2.76	2.53	2.37	2.25	2.17	2.10	2.04	1.99	1.95	1.92
	.01	7.08	4.98	4.13	3.65	3.34	3.12	2.95	2.82	2.72	2.63	2.56	2.50
120	.25	1.34	1.40	1.39	1.37	1.35	1.33	1.31	1.30	1.29	1.28	1.27	1.26
	.10	2.75	2.35	2.13	1.99	1.90	1.82	1.77	1.72	1.68	1.65	1.62	1.60
	.05	3.92	3.07	2.68	2.45	2.29	2.17	2.09	2.02	1.96	1.91	1.87	1.83
	.01	6.85	4.79	3.95	3.48	3.17	2.96	2.79	2.66	2.56	2.47	2.40	2.34
200	.25	1.33	1.39	1.38	1.36	1.34	1.32	1.31	1.29	1.28	1.27	1.26	1.25
	.10	2.73	2.33	2.11	1.97	1.88	1.80	1.75	1.70	1.66	1.63	1.60	1.57
	.05	3.89	3.04	2.65	2.42	2.26	2.14	2.06	1.98	1.93	1.88	1.84	1.80
	.01	6.76	4.71	3.88	3.41	3.11	2.89	2.73	2.60	2.50	2.41	2.34	2.27
∞	.25	1.32	1.39	1.37	1.35	1.33	1.31	1.29	1.28	1.27	1.25	1.24	1.24
	.10	2.71	2.30	2.08	1.94	1.85	1.77	1.72	1.67	1.63	1.60	1.57	1.55
	.05	3.84	3.00	2.60	2.37	2.21	2.10	2.01	1.94	1.88	1.83	1.79	1.75
	.01	6.63	4.61	3.78	3.32	3.02	2.80	2.64	2.51	2.41	2.32	2.25	2.18

15	20	24	30	40	50	60	100	120	200	500	∞	Pr	df for denominator N₂
1.36	1.34	1.33	1.32	1.31	1.31	1.30	1.30	1.30	1.29	1.29	1.28	.25	
1.81	1.76	1.73	1.70	1.67	1.65	1.64	1.61	1.60	1.59	1.58	1.57	.10	22
2.15	2.07	2.03	1.98	1.94	1.91	1.89	1.85	1.84	1.82	1.80	1.78	.05	
2.98	2.83	2.75	2.67	2.58	2.53	2.50	2.42	2.40	2.36	2.33	2.31	.01	
1.35	1.33	1.32	1.31	1.30	1.29	1.29	1.28	1.28	1.27	1.27	1.26	.25	
1.78	1.73	1.70	1.67	1.64	1.62	1.61	1.58	1.57	1.56	1.54	1.53	.10	24
2.11	2.03	1.98	1.94	1.89	1.86	1.84	1.80	1.79	1.77	1.75	1.73	.05	
2.89	2.74	2.66	2.58	2.49	2.44	2.40	2.33	2.31	2.27	2.24	2.21	.01	
1.34	1.32	1.31	1.30	1.29	1.28	1.28	1.26	1.26	1.26	1.25	1.25	.25	
1.76	1.71	1.68	1.65	1.61	1.59	1.58	1.55	1.54	1.53	1.51	1.50	.10	26
2.07	1.99	1.95	1.90	1.85	1.82	1.80	1.76	1.75	1.73	1.71	1.69	.05	
2.81	2.66	2.58	2.50	2.42	2.36	2.33	2.25	2.23	2.19	2.16	2.13	.01	
1.33	1.31	1.30	1.29	1.28	1.27	1.27	1.26	1.25	1.25	1.24	1.24	.25	
1.74	1.69	1.66	1.63	1.59	1.57	1.56	1.53	1.52	1.50	1.49	1.48	.10	28
2.04	1.96	1.91	1.87	1.82	1.79	1.77	1.73	1.71	1.69	1.67	1.65	.05	
2.75	2.60	2.52	2.44	2.35	2.30	2.26	2.19	2.17	2.13	2.09	2.06	.01	
1.32	1.30	1.29	1.28	1.27	1.26	1.26	1.25	1.24	1.24	1.23	1.23	.25	
1.72	1.67	1.64	1.61	1.57	1.55	1.54	1.51	1.50	1.48	1.47	1.46	.10	30
2.01	1.93	1.89	1.84	1.79	1.76	1.74	1.70	1.68	1.66	1.64	1.62	.05	
2.70	2.55	2.47	2.39	2.30	2.25	2.21	2.13	2.11	2.07	2.03	2.01	.01	
1.30	1.28	1.26	1.25	1.24	1.23	1.22	1.21	1.21	1.20	1.19	1.19	.25	
1.66	1.61	1.57	1.54	1.51	1.48	1.47	1.43	1.42	1.41	1.39	1.38	.10	40
1.92	1.84	1.79	1.74	1.69	1.66	1.64	1.59	1.58	1.55	1.53	1.51	.05	
2.52	2.37	2.29	2.20	2.11	2.06	2.02	1.94	1.92	1.87	1.83	1.80	.01	
1.27	1.25	1.24	1.22	1.21	1.20	1.19	1.17	1.17	1.16	1.15	1.15	.25	
1.60	1.54	1.51	1.48	1.44	1.41	1.40	1.36	1.35	1.33	1.31	1.29	.10	60
1.84	1.75	1.70	1.65	1.59	1.56	1.53	1.48	1.47	1.44	1.41	1.39	.05	
2.35	2.20	2.12	2.03	1.94	1.88	1.84	1.75	1.73	1.68	1.63	1.60	.01	
1.24	1.22	1.21	1.19	1.18	1.17	1.16	1.14	1.13	1.12	1.11	1.10	.25	
1.55	1.48	1.45	1.41	1.37	1.34	1.32	1.27	1.26	1.24	1.21	1.19	.10	120
1.75	1.66	1.61	1.55	1.50	1.46	1.43	1.37	1.35	1.32	1.28	1.25	.05	
2.19	2.03	1.95	1.86	1.76	1.70	1.66	1.56	1.53	1.48	1.42	1.38	.01	
1.23	1.21	1.20	1.18	1.16	1.14	1.12	1.11	1.10	1.09	1.08	1.06	.25	
1.52	1.46	1.42	1.38	1.34	1.31	1.28	1.24	1.22	1.20	1.17	1.14	.10	200
1.72	1.62	1.57	1.52	1.46	1.41	1.39	1.32	1.29	1.26	1.22	1.19	.05	
2.13	1.97	1.89	1.79	1.69	1.63	1.58	1.48	1.44	1.39	1.33	1.28	.01	
1.22	1.19	1.18	1.16	1.14	1.13	1.12	1.09	1.08	1.07	1.04	1.00	.25	
1.49	1.42	1.38	1.34	1.30	1.26	1.24	1.18	1.17	1.13	1.08	1.00	.10	∞
1.67	1.57	1.52	1.46	1.39	1.35	1.32	1.24	1.22	1.17	1.11	1.00	.05	
2.04	1.88	1.79	1.70	1.59	1.52	1.47	1.36	1.32	1.25	1.15	1.00	.01	

TABLE B-4
Upper Percentage Points of the χ^2 Distribution

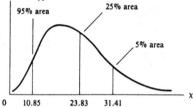

Example

Pr $(\chi^2 > 10.85) = 0.95$

Pr $(\chi^2 > 23.83) = 0.25$ for df = 20

Pr $(\chi^2 > 31.41) = 0.05$

Degrees of Freedom ＼ Pr	.995	.990	.975	.950	.900
1	$392704 \cdot 10^{-10}$	$157088 \cdot 10^{-9}$	$982069 \cdot 10^{-9}$	$393214 \cdot 10^{-8}$	.0157908
2	.0100251	.0201007	.0506356	.102587	.210720
3	.0717212	.114832	.215795	.351846	.584375
4	.206990	.297110	.484419	.710721	1.063623
5	.411740	.554300	.831211	1.145476	1.61031
6	.675727	.872085	1.237347	1.63539	2.20413
7	.989265	1.239043	1.68987	2.16735	2.83311
8	1.344419	1.646482	2.17973	2.73264	3.48954
9	1.734926	2.087912	2.70039	3.32511	4.16816
10	2.15585	2.55821	3.24697	3.94030	4.86518
11	2.60321	3.05347	3.81575	4.57481	5.57779
12	3.07382	3.57056	4.40379	5.22603	6.30380
13	3.56503	4.10691	5.00874	5.89186	7.04150
14	4.07468	4.66043	5.62872	6.57063	7.78953
15	4.60094	5.22935	6.26214	7.26094	8.54675
16	5.14224	5.81221	6.90766	7.96164	9.31223
17	5.69724	6.40776	7.56418	8.67176	10.0852
18	6.26481	7.01491	8.23075	9.39046	10.8649
19	6.84398	7.63273	8.90655	10.1170	11.6509
20	7.43386	8.26040	9.59083	10.8508	12.4426
21	8.03366	8.89720	10.28293	11.5913	13.2396
22	8.64272	9.54249	10.9823	12.3380	14.0415
23	9.26042	10.19567	11.6885	13.0905	14.8479
24	9.88623	10.8564	12.4011	13.8484	15.6587
25	10.5197	11.5240	13.1197	14.6114	16.4734
26	11.1603	12.1981	13.8439	15.3791	17.2919
27	11.8076	12.8786	14.5733	16.1513	18.1138
28	12.4613	13.5648	15.3079	16.9279	18.9392
29	13.1211	14.2565	16.0471	17.7083	19.7677
30	13.7867	14.9535	16.7908	18.4926	20.5992
40	20.7065	22.1643	24.4331	26.5093	29.0505
50	27.9907	29.7067	32.3574	34.7642	37.6886
60	35.5346	37.4848	40.4817	43.1879	46.4589
70	43.2752	45.4418	48.7576	51.7393	55.3290
80	51.1720	53.5400	57.1532	60.3915	64.2778
90	59.1963	61.7541	65.6466	69.1260	73.2912
100†	67.3276	70.0648	74.2219	77.9295	82.3581

† For df greater than 100 the expression: $\sqrt{2\chi^2} - \sqrt{(2k-1)} = Z$ follows the standardized normal distribution, where k represents the degrees of freedom.

.750	.500	.250	.100	.050	.025	.010	.005
.1015308	.454937	1.32330	2.70554	3.84146	5.02389	6.63490	7.87944
.575364	1.38629	2.77259	4.60517	5.99147	7.37776	9.21034	10.5966
1.212534	2.36597	4.10835	6.25139	7.81473	9.34840	11.3449	12.8381
1.92255	3.35670	5.38527	7.77944	9.48773	11.1433	13.2767	14.8602
2.67460	4.35146	6.62568	9.23635	11.0705	12.8325	15.0863	16.7496
3.45460	5.34812	7.84080	10.6446	12.5916	14.4494	16.8119	18.5476
4.25485	6.34581	9.03715	12.0170	14.0671	16.0128	18.4753	20.2777
5.07064	7.34412	10.2188	13.3616	15.5073	17.5346	-20.0902	21.9550
5.89883	8.34283	11.3887	14.6837	16.9190	19.0228	21.6660	23.5893
6.73720	9.34182	12.5489	15.9871	18.3070	20.4831	23.2093	25.1882
7.58412	10.3410	13.7007	17.2750	19.6751	21.9200	24.7250	26.7569
8.43842	11.3403	14.8454	18.5494	21.0261	23.3367	26.2170	28.2995
9.29906	12.3398	15.9839	19.8119	22.3621	24.7356	27.6883	29.8194
10.1653	13.3393	17.1170	21.0642	23.6848	26.1190	29.1413	31.3193
11.0365	14.3389	18.2451	22.3072	24.9958	27.4884	30.5779	32.8013
11.9122	15.3385	19.3688	23.5418	26.2962	28.8454	31.9999	34.2672
12.7919	16.3381	20.4887	24.7690	27.5871	30.1910	33.4087	35.7185
13.6753	17.3379	21.6049	25.9894	28.8693	31.5264	34.8053	37.1564
14.5620	18.3376	22.7178	27.2036	30.1435	32.8523	36.1908	38.5822
15.4518	19.3374	23.8277	28.4120	31.4104	34.1696	37.5662	39.9968
16.3444	20.3372	24.9348	29.6151	32.6705	35.4789	38.9321	41.4010
17.2396	21.3370	26.0393	30.8133	33.9244	36.7807	40.2894	42.7956
18.1373	22.3369	27.1413	32.0069	35.1725	38.0757	41.6384	44.1813
19.0372	23.3367	28.2412	33.1963	36.4151	39.3641	42.9798	45.5585
19.9393	24.3366	29.3389	34.3816	37.6525	40.6465	44.3141	46.9278
20.8434	25.3364	30.4345	35.5631	38.8852	41.9232	45.6417	48.2899
21.7494	26.3363	31.5284	36.7412	40.1133	43.1944	46.9630	49.6449
22.6572	27.3363	32.6205	37.9159	41.3372	44.4607	48.2782	50.9933
23.5666	28.3362	33.7109	39.0875	42.5569	45.7222	49.5879	52.3356
24.4776	29.3360	34.7998	40.2560	43.7729	46.9792	50.8922	53.6720
33.6603	39.3354	45.6160	51.8050	55.7585	59.3417	63.6907	66.7659
42.9421	49.3349	56.3336	63.1671	67.5048	71.4202	76.1539	79.4900
52.2938	59.3347	66.9814	74.3970	79.0819	83.2976	88.3794	91.9517
61.6983	69.3344	77.5766	85.5271	90.5312	95.0231	100.425	104.215
71.1445	79.3343	88.1303	96.5782	101.879	106.629	112.329	116.321
80.6247	89.3342	98.6499	107.565	113.145	118.136	124.116	128.299
90.1332	99.3341	109.141	118.498	124.342	129.561	135.807	140.169

REFERENCES

AKAIKE, H. (1969) "Fitting autoregressions for prediction." *Annals of the Institute of Statistical Mathematics* 21: 243-247.

AKAIKE, H. (1970a) "Autoregressive model fitting for control." *Annals of the Institute of Statistical Mathematics* 22: 163-180.

AKAIKE, H. (1970b) "A fundamental relation between predictor identification and power spectrum estimation." *Annals of the Institute of Statistical Mathematics* 22: 219-223.

AKAIKE, H. (1973) "Information theory and an extension of the maximum likelihood principle," in B. N. Petrov and F. Csaki (eds.), *2nd International Symposium on Information Theory*, pp. 267-281. Budapest: Akademiai Kiado.

AKAIKE, H. (1979) "A Bayesian extension of the minimum AIC procedure." *Biometrika* 66: 237-42.

ANDERSON, T. W. (1963) "Determination of the order of dependence in normally distributed time series," in M. Rosenblatt (ed.), *Proceedings of the Symposium of Time Series Analysis*, pp. 425-446. New York: John Wiley.

ANDERSON, T. W. (1971) *The Statistical Analysis of Time Series*. New York: John Wiley.

BARTELS, R. (1982) "The rank version of Von Neumann's ratio test for randomness." *Journal of the American Statistical Association* 77: 40-46.

BAUMOL, W. J., and BENHABIB, J. (1989) "Chaos: Significance, mechanism, and economic applications." *Journal of Economic Perspectives* 3: 77-105.

BENHABIB, J., and NISHIMURA, K. (1979) "The Hopf bifurcation and the existence and stability of closed orbits in multisector models of optimum economic growth." *Journal of Economic Theory* 21: 421-444.

BERA, A. K., and JARQUE, C. M. (1981) "An efficient large sample test for normality of observations and regression residuals." Working Paper in Econometrics No. 40, Australian National University, Canberra.

BHANSALI, R. J., and DOWNHAM, D. J. (1977) "Some properties of autoregressive model selected by a generalization of Akaike's EPF criterion." *Biometrika* 64: 547-551.

BHARGAVA, A. S. (1986) "On the theory of testing for unit roots in observed time series." *Review of Economic Studies* 53: 369-384.

BOLLERSLEV, T. (1986) "Generalized autoregressive conditional heteroscedasticity." *Journal of Econometrics* 31: 307-327.

BOLLERSLEV, T. (1988) "On the correlation structure for the generalized autoregressive conditional heteroscedastic process." *Journal of Time Series Analysis* 9: 121-131.

BOX, G. E. P., and JENKINS, G. M. (1976) *Time Series Analysis: Forecasting and Control*. Revised edition. San Francisco: Holden-Day.

BOX, G. E. P., and PIERCE, D. A. (1970) "Distribution of residual autocorrelations in autoregressive moving average time series models." *Journal of the American Statistical Association* 65: 1509-1526.

BROCK, W. A. (1987) "Notes on nuisance parameter problems in BDS type tests for IID." Working Paper, revised November 1989. University of Wisconsin, Madison.

BROCK, W. A., and BAEK, E. G. (1991) "Some theory of statistical inference for nonlinear science." *Review of Economic Studies* 58: 697-716.

BROCK, W. A., DECHERT, W., and SCHEINKMAN, J. (1986) "A test of independence based on the correlation dimension." *SSRI Report #8702*, Department of Economics, University of Wisconsin, University of Houston, and University of Chicago.

BROCK, W. A., HSIEH, D. A., and LEBARRON, B. (1991) *Nonlinear Dynamics, Chaos and Instability*. Cambridge: MIT Press.

BROCKETT, R. W. (1976) "Volterra series and geometric control theory." *Automatica* 12: 167-176.

CHAN, K. S., and TONG, H. (1986) "On tests for nonlinearity in time series analysis." *Journal of Forecasting* 5: 217-218.

COCHRANE, J. H. (1988) "How big is the random walk in GNP?" *Journal of Political Economy* 96: 893-920.

COCHRANE, J. H. (1991) "A critque of the application of unit root tests." *Journal of Economic Dynamics and Control* 15: 275-284.

CONOVER, W. J. (1980) *Practical Nonparametric Statistics*. New York: John Wiley.

D'AGOSTINO, R. B., and STEPHENS, M. A. (1986) *Goodness-of-fit techniques*. New York: Marcel Dekker.

DE GOOIJER, J. G., ABRAHAM, B., GOULD, A., and ROBINSON, L. (1985) "Methods for determining the order of an autoregressive-moving average process: A survey." *International Statistical Review* 53: 301-329.

DE GOOIJER, J. G., and KUMAR, K. (1992) "Some recent developments in non-linear time series modeling, testing and forecasting." *International Journal of Forecasting* 8: 135-156.

DICKEY, D. A., BELL, W. R., and MILLER, R. B. (1986) "Unit roots in time series models: Tests and implications." *American Statistician* 40: 55-70.

DICKEY, D. A., and FULLER, W. A. (1979) "Distribution of the estimators for autoregressive time series with a unit root." *Journal of the American Statistical Association* 74: 427-431.

DICKEY, D. A., and FULLER, W. A. (1981) "Likelihood ratio tests for autoregressive time series with a unit root." *Econometrica* 49: 1057-1072.

DICKEY, D. A., and PANTULA, S. G. (1987) "Determining the order of differencing in autoregressive processes." *Journal of Business and Economic Statistics* 5: 455-462.

DIEBOLD, F. X., and NERLOVE, M. (1989) "Unit roots in economic time series: A selective survey," in T. B. Fomby and G. F. Rhodes (eds.), *Advances in Econometrics: Cointegration, Spurious Regressions, and Unit Roots*. Greenwich, CT: JAI Press.

ENGLE, R. (1982) "Autoregressive conditional heteroscedasticity with estimates of the variance of the U.K. inflation." *Econometrica* 50: 987-1007.

ENGLE, R. F., and BOLLERSLEV, T. (1986) "Modelling the persistence of conditional variances." *Econometric Reviews* 5,1: 1-50.

ENGLE, R. F., LILIEN, D. M., and ROBBINS, R. P. (1987) "Estimating time varying risk premia in the term structure: The ARCH-M model." *Econometrica* 55: 391-408.

EVANS, G. B. A., and SAVIN, N. E. (1981) "Testing for unit roots: 1." *Econometrica* 49: 753-779.

FULLER, W. A. (1976) *An Introduction to Statistical Time Series*. New York: John Wiley.

GABR, M. M., and SUBBA RAO, T. (1981) "The estimation and prediction of subset bilinear time series models with applications." *Journal of Time Series Analysis* 2: 155-171.

92

GEWEKE, J. F., and MEESE, R. A. (1981) "Estimating regression models of finite but unknown order." *International Economic Review* 22: 55-70.

GLEICK, J. (1987) *Chaos: Making a New Science.* New York: Viking Penguin.

GODFREY, L. G. (1979) "Testing the adequacy of a time series model." *Biometrika* 66: 67-72.

GRANGER, C. W. J. (1991) "Developments in the nonlinear analysis of economic series." *Scandanavian Journal of Economics* 93: 263-276.

GRANGER, C. W. J., and ANDERSEN, A. P. (1978) *An Introduction to Bilinear Time Series Models.* Gottingen, The Netherlands: Vandenhoek & Ruprechet.

GRANGER, C. W. J., and HUGHES, A. D. (1971) "A new look at some old data: The Beveridge wheat price series." *Journal of the Royal Statistical Society A* 134: 413-428.

GRANGER, C. W. J., KING, M., and WHITE, H. (1992) "Comments on testing economic theories and the use of model selection criteria." Discussion Paper 92-18, Department of Economics, University of California, San Diego.

GRANGER, C. W. J., and NEWBOLD, P. (1986) *Forecasting Economic Time Series.* 2nd edition. New York: Academic Press.

HAGGAN, V., and OZAKI, T. (1981) "Modelling non-linear random vibrations using an amplitude-dependent autoregressive time series model." *Biometrika* 68: 184-189.

HANNAN, E. J. (1980) "The estimation of the order of an ARMA process." *American Statistician* 8: 1071-1081.

HANNAN, E. J., and QUINN, B. G. (1979) "The determination of the order of an autoregression." *Journal of the Royal Statistical Society* Series B 41: 190-195.

HINICH, M. (1982) "Testing for gaussinity and linearity of a stationary time series." *Journal of Time Series Analysis* 3: 169-176.

HOWEREY, E. P. (1978) "Comments on estimating structural models of seasonality by R. F. Engle," *Seasonal Analysis of Economic Time Series,* ed. A. Zellner (pp. 298-302). Washington, DC: U.S. Department of Commerce, Bureau of the Census.

HSIEH, D. A. (1989) "Testing for nonlinear dependence in daily foreign exchange rates." *Journal of Business* 62: 339-368.

KEENAN, D. M. (1985) "A Tukey nonadditivity-type test for time series non-linearity." *Biometrika* 72: 39-44.

KENDALL, M. G., and STUART, M. A. (1961) *The Advanced Theory of Statistics.* Volume 2. London: Griffin.

KUMAR, K. (1986) "On the identification of some bilinear time series models." *Journal of Time Series Analysis* 7: 117-122.

LABYS, W. C., and GRANGER, C. W. J. (1970) *Speculation, Hedging and Commodity Price Forecasts.* Lexington, MA: Heath Lexington.

LABYS, W. C., MURCIA, V., and TERRAZA, M. (1991) "Progres Econometriques et Series Temporelles." Working Paper, Center for Industrial Econometrics, University of Montpellier I.

LEE, T. H., WHITE, H., and GRANGER, C. W. J. (1993) "Testing for neglected nonlinearity in time series models." *Journal of Econometrics* 56: 289-296.

LJUNG, G. M., and BOX, G. E. P. (1978) "On a measure of the lack of fit in time series models." *Biometrika* 65: 297-303.

LUUKKONEN, R., SAIKKONEN, P., and TERASVIRTA, T. (1988) "Testing linearity against smooth transition autoregressive models." *Biometrika* 75: 491-500.

MacKINNON, J. G. (1990) "Critical values for co-integration tests." Working Paper. University of California, San Diego.

MacKUEN, M., ERIKSON, R. S., and STIMSON, J. A. (1992) "Moving attractors and character of partisan change." *American Political Science Review* 86: 100-110.

McCLAVE, J. (1975) "Subset autoregressive." *Technometrics* 17: 213-220.

McLEARY, R., and HAY, R. (1980) *Applied Time Series Analysis*. Beverly Hills, CA: Sage.

McLEOD, A. I., and LI, W. K. (1983) "Diagnostic checking ARMA time series models using squared-residuals autocorrelations." *Journal of Time Series Analysis* 4: 269-273.

MicroTSP 7.0. (1992) Quantitative Micro Software. Irvine, CA.

MILLS, T. C. (1990) *Time Series Techniques for Economists*. Cambridge: Cambridge University Press.

OZAKI, T. (1980) "Non-linear time series models for non-linear random vibrations." *Journal of Applied Probability* 17: 84-93.

PARZEN, E. (1974) "Some recent advances in time series modeling." *IEEE Transactions on Automatic Control* AC-19: 723-730.

PEARSON, E. S., and HARTLEY, H. O. (1966) *Biometrika Tables for Statisticians* (vol. 1, 3rd ed.). New York: Cambridge University Press.

PETRUCCELLI, J. D., and DAVIES, N. (1986) "A portmanteau test for self exciting threshold autoregressive-type non-linearity in time series." *Biometrika* 73: 687-694.

PHILLIPS, P. C. B. (1987) "Time series with a unit root." *Econometrica* 55: 277-301.

PHILLIPS, P. C. B., and PERRON, P. (1986) "Testing for a unit root in time series regressions." DP No. 781. Cowles Foundation, Yale University, New Haven.

PRIESTLEY, M. B. (1980) "State-dependent models: A general approach to non-linear time series analysis." *Journal of Time Series Analysis* 1: 47-71.

PRIESTLEY, M. B. (1981) *Spectral Analysis and Time Series*. San Diego, CA: Academic Press.

PRIESTLEY, M. B. (1988) *Non-linear and non-stationary time series analysis*. New York: Academic Press.

RAMSEY, J. B. (1969) "Tests for specification error in classical linear least squares regression analysis." *Journal of the Royal Statistical Society B* 31: 250-271.

RATS 3.0, VAR Econometrics, Minneapolis, MN.

SAID, S. E., and DICKEY, D. A. (1984) "Testing for unit roots in autoregressive moving-average models with unknown order." *Biometrika* 71: 599-607.

SAID, S. E., and DICKEY, D. A. (1985) "Hypothesis testing in ARIMA (p,1,q) models." *Journal of the American Statistical Association* 80: 369-374.

SCHWARZ, G. (1978) "Estimating the dimension of a model." *Annals of Statistics* 6: 461-464.

SCHWERT, G. W. (1987) "Effects of model specification on tests for unit roots in macroeconomic data." *Journal of Monetary Economics* 20: 73-103.

SHAZAM. *User's Reference Manual*. Version 6.2. New York: McGraw-Hill.

STIMSON, J. A. (1991) *Public Opinion in America: Moods, Cycles and Swings*. Boulder, CO: Westview.

SUBBA RAO, T. (1981a) "A cumulative sum test for detecting change in time series." *International Journal of Control* 34: 285-293.

SUBBA RAO, T. (1981b) "On the theory of bilinear time series models." *Journal of Royal Statistical Society B* 43: 244-255.

SUBBA RAO, T., and GABR, M. M. (1980) "A test for linearity of stationary time series." *Journal of Time Series Analysis* 1: 145-148.

SUBBA RAO, T., and GABR, M. M. (1984) *An Introduction to Bispectral Analysis and Bilinear Time Series Models*. Lecture Notes in Statistics Vol. 24. Berlin: Springer-Verlag.

TAKENS, F. (1980) "Detecting strange attractors in turbulence," in D. Rand and B. S. Yound (eds.), *Dynamical Systems and Turbulence*. Lecture Notes in Mathematics 898. Berlin: Springer-Verlag.

TERRAZA, M. (1980) "Etude conjoncturelle de la consommation taxée des vins d'appellation d'origine contrôlée par la methode Demter." *Economie Appliquée* 3-4: 787-820.

TERRAZA, M. (1981) "Previsions a court terme des series temporelles economiques— l'estimation des modeles ARIMA par l'algorithme de la plus forte pente." *Journal de la Societe de Statistique de Paris* 3: 143-161.

TONG, H. (1979) "A note on local equivalence of two recent approaches to autoregressive order determination." *International Journal of Control* 29: 441-446.

TONG, H. (1983) *Threshold Models in Non-Linear Time Series Analysis*. Berlin: Springer-Verlag.

TONG, H., and LIM, K. S. (1980) "Threshold autoregression, limit cycles and cyclical data." *Journal of the Royal Statistical Society* Series B 42: 245-292.

TSAY, R. S. (1986) "Nonlinearity test for time series." *Biometrika* 73: 461-466.

WALLIS, K. F. (1987) "Time series analysis of bounded economic variables." *Journal of Time Series Analysis* 8: 115-123.

WEISS, A. A. (1984) "ARMA models with ARCH errors." *Journal of Time Series Analysis* 5: 129-143.

ABOUT THE AUTHORS

JEFF B. CROMWELL is a Research Fellow at the Institute for Labor Studies and a Ph.D. candidate in the Natural Resource Economics program in the College of Agriculture and Forestry at West Virginia University. His research interests have been in the areas of nonlinear time series analysis and resource modeling. His teaching experience includes West Virginia University, Edinboro University of Pennsylvania, California University of Pennsylvania, and the Inter-University Consortium for Political and Social Science Research at the University of Michigan. He has recently published an article on multivariate time series methodology in the *International Regional Science Review*.

WALTER C. LABYS is Benedum Distinguished Scholar and Professor of Resource Economics in the College of Agriculture and Forestry at West Virginia University. He recently served as Visiting Professor at the Center for Industrial Econometrics at the University of Montpellier, where he collaborated with Michel Terraza on the present text. His interest in time series analysis began before 1970 when he copublished his dissertation, *Speculation, Hedging and Commodity Price Forecasts*. Since then he has published articles on the application of time series methods to international commodity markets in journals such as the *Oxford Bulletin of Economics and Statistics, Applied Economics, Journal of Futures Markets, Journal of Development Studies, Resources Policy, Journal of Policy Modeling, Energy Economics*, and *Weltwirtschaftliches Archives*. He is currently involved with the time series analysis of international commodity markets, inventories, and prices.

MICHEL TERRAZA is a Senior Lecturer in the Faculty of Economics at the University of Montpellier I. He is founder of the Centre d'Econometrie Pour l'Enterprise (C.E.P.E.) at the University of Montpellier I, that is, the Center for Industrial Econometrics. He also is involved with the Research Group on Statistics and Mathematics at the École d'Ingenieurs des Mines d'Ales. His research interests have been in time series analysis, particularly spectral analysis, univariate and multivariate analysis with applications to a number of industries. His teaching interests include the development of undergraduate, graduate, and short courses on time series analysis. The latter courses have been taught at the management and public policy level. Recent research includes collaboration with Walter Labys on the present monograph and on other industry modeling projects. In addition to serving as editor for the C.E.P.E. Working Paper Series, he also has consulted with businesses and governments on economic conditions and industry development.

Quantitative Applications in the Social Sciences

A SAGE UNIVERSITY PAPERS SERIES

$9.95 each

SAGE PUBLICATIONS, INC.
P.O. BOX 5084
THOUSAND OAKS, CALIFORNIA 91359-9924